Crafting Masculine Selves

Crafting Masculine Selves

Culture, War, and Psychodynamics in Afghanistan

ANDREA CHIOVENDA

OXFORD
UNIVERSITY PRESS

Oxford University Press is a department of the University of Oxford. It furthers
the University's objective of excellence in research, scholarship, and education
by publishing worldwide. Oxford is a registered trade mark of Oxford University
Press in the UK and certain other countries.

Published in the United States of America by Oxford University Press
198 Madison Avenue, New York, NY 10016, United States of America.

CIP data is on file at the Library of Congress
ISBN 978–0–19–007355–8

1 3 5 7 9 8 6 4 2

Printed by Integrated Books International, United States of America

Contents

Acknowledgments

This book would not have been written and published without the participation and help of many people and institutions, on multiple levels.

I need first to thank my anthropology faculty and department at Boston University (BU), where I took my first step in anthropology as a graduate student. Thomas Barfield and Charles Lindholm, in particular, were the ones with whom I more closely developed my research about Afghanistan, Pashtun people and psychological anthropology at BU. For that I am most grateful. They wanted me at BU, led my study and research in my first years, and have remained good friends with myself and my family. Robert Weller, Parker Shipton, Nancy Smith-Hefner, Robert Hefner, Shahla Haeri, Jenny White, Matt Cartmill and Jeremy Da Silvaare scholars from whom I learned very much as well, and who have strongly contributed to my intellectual formation. Joanna Davidson and Kimberly Arkin, in their very first years of teaching at BU when I started my coursework, have been an important point of reference for me, even though I was not directly one of their students. I want to make a special mention to Adam Kuper, who co-taught a dissertation writing seminar at BU. His advice was important for my doctoral manuscript, and the friendship we built on that seminar I still cherish and nurture. Equally I wish to thank Nazif Shahrani, at Indiana University, of whose scholarship and personal experience of Afghanistan I have greatly benefited over the years.

Yet my intellectual trajectory at BU soon brought me into a parallel institutional environment—Harvard University. When I started my graduate studies, Robert LeVine had just stopped teaching at BU after retiring as Emeritus from a long career at Harvard, and we, luckily for me, met at a department celebration at BU. Our common interest in psychoanalysis sparked an academic collaboration and sincere friendship that lasts to this day. I cannot even start to measure the impact that our monthly conversations over lunch had on the development of my ideas and the integration of psychoanalytic models into the practice and theory of psychological anthropology. Bob was one of the main readers of my dissertation at BU.

Bob also initially introduced me to the Friday Morning Seminar (FMS) in Culture, Psychiatry, and Global Mental Health at Harvard University, a long-standing seminar series created by Byron Good, Mary-Jo DelVecchio Good, and Arthur Kleinman in 1984. That's where, in 2010, I first met Byron and Mary-Jo. The seminar proved to be a unique intellectual breeding ground for ideas, analysis, and topics that were close to my specific interests, and I never stopped attending it since. Many years later, today I am honored to be the "manager," for lack of a better term, of the FMS, which is now chaired by Mary-Jo, Byron, and Michael Fischer (whom I want to thank for all the times he let me pick his brain). My relationship with Byron and Mary-Jo has morphed through the years into a mentoring one, a professional one (as their research assistant), and eventually one of close friendship, for which I am very grateful. Over the past four years, at the Department of Global Health and Social Medicine, Harvard Medical School, we have taught together, created syllabi and developed courses together, and enjoyed the weekly discovery of new ideas at FMS, in an atmosphere of quotidian intellectual exchange and enrichment that has immensely benefited my work and personal growth. Like Bob LeVine, Byron was a reader for my dissertation at BU.

I wish certainly to thank all the people who have worked with me in Afghanistan, without whom, again, this book would not have seen the light of the day; and all the friends at the American Institute of Afghanistan Studies in Kabul (directed by Thomas Barfield for many years), who have helped me and my wife over the many years of visits and longer permanence in town: Zafar, Sultan, the late Saeed Niazi, Jibrael Amin, Rohullah Amin, Palahwan and his family have all been crucial in many occasions to help us logistically and emotionally. A special mention I want to dedicate to Omar Sharifi, an Afghan physician whom I met as the field director in Kabul of AIAS, and found again as a colleague in the PhD program in anthropology at BU after my fieldwork. Now happily graduated, he was the companion of countless and endless conversations in both Kabul and Boston about Afghanistan, anthropology, and much more.

I, of course, owe an unrepayable debt to all those Afghans who have closely collaborated with me, helped me in my fieldwork research, and eventually became the subjects of it in the south-east provinces of the country. I will not name any of them by their real name here for obvious reasons of security. Yet for all of them who will read this book, I want to reiterate the deep gratitude and appreciation I feel toward them.

I also want to mention here those Pashtun friends I have in Peshawar, Pakistan, who supported me in the very early stages of my research with Pashtun people, and in the United States, who were competent, patient, and caring language teachers so that in the end I managed to get to Afghanistan with a good knowledge of spoken and written Pashto language. I will not name any of them here for the same reasons as above, but I am likewise deeply indebted to them.

I want to thank my family and my parents. The unwavering support that my parents, Romano and Grazia Maria, have given me through the years (and not only in the material sense) proved indispensable for the continuation of my studies after I went back to graduate school for my "second career." I thank my wife, Melissa, an anthropologist in her own right, for bearing my worst moments without giving up, and for being the source of invaluable inputs and ideas during our animated conversations. And our daughter, Aurelia, whose contagious smile has always cleared up the most overcast skies of Boston.

I wish to acknowledge and sincerely thank the anonymous reviewers who read the manuscript of this book and provided me with crucial comments and suggestions. Likewise, I warmly thank the past and present editors at Oxford University Press—Angela Chnapko, Alexcee Bechthold, and Anne Dellinger—who proved to be an extremely professional, competent, and friendly team, as well as Preetham Raj at Newgen Knowledge Works, whose editing and proofreading efforts were just as much impeccable and friendly.

The many colleagues and senior scholars with whom I had the chance to discuss my work over the years are too many to acknowledge here properly and in full, but I wish to let them all know that I am grateful of all the inputs, feedback, and inspiration they have provided me.

And last but not least, I certainly need to thank our Afghan dog, Bruno, for supporting me during some difficult moments in my fieldwork in Jalalabad, and our American dog, Nero, for lighting up (alongside Bruno) many of my days in Boston.

While I deeply acknowledge all the contributions to this book I have highlighted so far, I take full and sole responsibility for all the shortcomings and issues that it will display.

Introduction

Prologue

During one of our conversations, Rohullah recounted the following story:

ROHULLAH—*My older brother Zair was coming back home from work that day, and met one of the sons of our neighbor in front of their house gate. He started talking to the guy, asking him why they were being so difficult and disrespectful about this problem* [their neighbor had recently added two stories to his house, from which the courtyard of Rohullah's family's house could be seen, compromising the privacy of Rohullah's family's female members]. *Then, I don't know exactly what happened, or what they said to each other. I heard Zair screaming outside, and the other guy screaming back at him. You know how Zair is, he does not have much patience when he thinks that he is being disrespected, he gets upset quickly. By the time I got in the street they were hitting each other. I called out Iqbal* [a younger brother] *and my father, and when they arrived there were two or three other people from the other family outside. They had sticks with them, and when my father tried to separate Zair and the other guy, they intervened and hit my father on the head. When I saw my father being hit, I completely lost myself. I went back to my courtyard, picked up an ax, and went outside. I started swinging the ax, and I think I got someone in the arm. Zair and Iqbal had gotten ahold of a stick too, and were fighting with the other guys. Soon after, other people from the neighborhood came to the scene and put themselves between us and the other family. So we stopped fighting. At least one of them ended up at the hospital, I believe. My father later went to the police and denounced what had happened, so that we would be on the safe side. We have yet to reach a solution to the problem of the wall, and to solve the issue of the enmity* [dukhmani] *between families that has started with this fight.*

ANDREA—*How do you feel about the fight?*

ROHULLAH—*Well, that was good ghairat* [the masculine virtue of defending one's family's respectability]. *I saw my father being hit, what else could I do?*

Crafting Masculine Selves. Andrea Chiovenda, Oxford University Press (2020). © Oxford University Press.
DOI: 10.1093/oso/9780190073558.001.0001

And also, these people, they are doing something really wrong. . . . I mean, they are not respecting the parda [privacy] of the women of our family. What else could have we done? Sooner or later we would have ended up fighting, one way or another. I think that it is good when I manage to be aggressive in a situation that really requires it. This is good ghairat. I am proud of my ghairat in situations like these. It makes me feel like a real Pashtun.

This incident took place in mid-2012 in a middle-class, Pashtun-majority neighborhood of Kabul, home to many rural Pashtuns who relocated to the city from the eastern provinces of Afghanistan (where I conducted my fieldwork research). Rohullah is one of the close informants (see chapter 3) with whom I worked on a regular basis while in the field. As we will see, he is a young man in his late twenties (in 2012), is married with a daughter, has a college degree, and lives now in Kabul. He hails from a middle-class and "intellectual" Pashtun family. The incident he recalled in this passage represents not an "exception" within the Pashtun sociocultural context, and certainly not an event one would expect to take place only in a rural environment, or among the less-educated, more "traditionalist" strata of the population. The moral values, emotional affects, and behavioral ethics that emerge from this short passage pervade the life of any Pashtun man, wittingly or unwittingly, regardless of social and economic background. During my fieldwork I discovered that life (particularly psychological life) in such an environment represented a daily challenge for most of my informants, and presented constant interpersonal frictions.

This book, which is the result of over eighteen months of fieldwork research, carried out between 2009 and 2013 mainly in Nangarhar, a southeastern province of Afghanistan, will not focus on male Afghan Pashtuns' "culture" per se, or Pashtun "social structure," or even political dynamics. Rather, it represents a psychological investigation of individual lives lived within these interconnected contexts. Though I will later sketch briefly the features that ethnographic literature has attributed to Pashtun society, I will let the reader become gradually acquainted with the reality of these features as experienced by the people I engage in the following ethnographic chapters, without aprioristic assumptions. This is also the way in which I have incrementally, and experientially, come to know the sociocultural environment in which I worked.

Indeed, one of the central arguments in this book is that the study of cultural idioms and public behaviors, taken at face value, cannot tell the whole story about individuals living in any given social group: a psychological investigation of individual subjectivity is necessary to understand the personal meanings given to such idioms and behaviors, and the impact they have on one's inner dynamics. Awareness of the private conflicts possibly entailed by these subjective encounters yields a better understanding of the whole sociocultural context, and of the possibility for cultural and social change within it. The cultural lens through which I chose to look at these encounters is masculinity, a prominent idiom in a strongly androcentric cultural milieu, such as the Pashtuns'.

A psychodynamic framework of investigation

To explore one's subjectivity means to try to shed light onto thought processes, emotional reactions, and unconscious dynamics of the individual under observation. Indeed, a crucial aspect of such endeavor, which I develop in this book, lies in paying attention

> To that which is not said overtly, to that which is unspeakable and unspoken, that which appears at the margin of formal speech and everyday presentations of self, manifest in the Imaginary, in dissociated spaces and individual dream time, and in traces of the apparently forgotten, coded in esoteric symbol productions aimed at hiding as well as revealing. (Good B. 2012b: 519, emphasis in original)

In turn, doing so from an anthropological perspective means assuming that subjectivity is not an aspect of the "self" that exists in a vacuum, independent of the social and cultural contexts in which the individual is immersed, but rather that the workings of the mind must be read and interpreted through the lens of the sociocultural environment in which such mind developed. Even more crucially, it means paying attention to how psychic dynamics are infused with cultural meaning, and in so doing operationalized accordingly to obtain public behavior in a specific sociocultural setting.

Psychological and psychoanalytic literature has been acknowledging, and critiquing, the "myth of the isolated mind" for about two decades now (see Stolorow and Atwood 1994: 233–250). However, it is still uncommon among psychoanalysts to fully conceive of what affects the human psyche beyond interpersonal dyadic relations (be it between patient and analyst, or between the patient and a third person in daily life), or triadic oedipal constellations at best (for a remarkable exception, see Dalal 2002, 2006).[1,2] While it is true that within psychoanalytic and psychological circles there exists a "resistance" (to use an apt term in this case) to recognize the role of cultural and social milieus in shaping the inner dynamics of each individual, it is also true that within anthropological circles one encounters a similar resistance to the acknowledgment of the importance of individual psychic dynamics for a fuller understanding of broader social phenomena.[3]

My work aims at reconciling these two seemingly mutually exclusive realms of investigation. My primary goal is that of elucidating the psychological processes my informants go through in their daily lives, and understanding the subjective meanings they give to their private and social experiences. The relevance of this main objective to a broader picture of

[1] Certainly, psychoanalysts-cum-anthropologists such as George Devereux and Erik Erikson expanded psychological investigation fully into the realm of cultural and social spheres already in the 1940s and 1950s, yet their approach to the study of mind was still strongly informed by a Freudian metapsychology and its assumptions, which curtailed their appreciation of the depth of culture's reach into the individual's mental and emotional processes. Robert Levy, also a psychoanalyst/anthropologist, moved away from psychodynamic research, and chose to delve into what became later known as "ethnopsychology" (Levy 1973). Two "full-time" psychoanalysts who decided to carry out outright ethnographic work, in the 1950s and early 1960s, are G. M. Carstairs and Wulf Sachs. Carstairs (1958) remained rather anchored to the school of "culture and personality" studies opened by Ruth Benedict and Margaret Mead (with all the pitfalls that it entailed), while Sachs (1968 [1947]) produced an extremely successful experiment at a culturally informed psychological study of one single individual (which, maybe for this reason, has unfortunately remained peripheral to the interest of professional anthropologists). More-recent attempts at bridging the gap between psychoanalysis and sociocultural context, carried out by professional psychoanalysts, are Erich Fromm's (with anthropologist Michael Maccoby. Fromm and Maccoby 1970), and Alan Roland's (Roland 1988). Neither of them broke much ground among anthropologists, for various reasons (not least of which was parochialism), in spite of many interesting and insightful aspects that both works presented. A professional anthropologist, fully trained in psychoanalysis at the Chicago Psychoanalytic Institute, under Heinz Kohut, is Robert LeVine, who is still very much active today. In 1973 he published a groundbreaking volume (revised in 1982) in the theory and practice of psychoanalytic ethnography, which has widely influenced the following generations of psychological anthropologists, including myself (one of his students, Waud Kracke, was very successful in the production of a psychoanalytically oriented ethnography. Kracke 1978). More recently, two major figures in anthropology with full training in psychoanalysis have been Katherine Pratt Ewing and Douglas Hollan (the latter being also a practicing psychoanalyst). Their clinically-oriented academic articles and broader book-length ethnographies have been a crucial point of reference and inspiration for the development of this book (Ewing 1987, 1990, 1991, 1992, 1997, 2008; Hollan 1992, 2000, 2001, 2016; Hollan and Wellenkamp 1994, 1996).

[2] The contemporary field of developmental psychology, especially in its more specific branch dedicated to "attachment theory" (derived from the initial propositions of psychoanalyst John Bowlby in the 1960s, and his student Mary Ainsworth later), is particularly impervious to the acknowledgment of the importance of the cultural and social context in which the child develops his or her first mental functions. Two edited volumes have been recently published to counter this strong and widespread attitude in developmental psychology, one mainly authored by anthropologists (Quinn and Mageo 2013), the other mainly by psychologists (Otto and Keller 2014).

[3] Indeed, Karin Norman and Robert LeVine wrote that "Psychologists have generally been as immune to a cultural approach as anthropologists are to 'psychologizing'" (LeVine and Norman 2001: 86).

social dynamics lies in the realization that, as Norbert Elias suggested long ago, every society is a "society of individuals" (Elias 1991 [1939]). Far from constituting a "superorganic" structure with its rules and impalpable reality (à la Kroeber), or a set of fixed social roles/functions embodied in turn by different individuals (à la Radcliffe-Brown), I believe with Elias that every society is in fact represented by the mutual relations of *interdependence* between its members—the individuals (Elias 1994 [1968]: 225). Investigating the way each individual privately interprets and inhabits his or her sociocultural environment means understanding those very relations of interdependence on which society is premised.[4] Edward Sapir, a contemporary of Norbert Elias, tackled similar issues as Elias, only starting from the opposite side of the continuum—the individual. While he wrote that "the individual in isolation from society is a psychological fiction" (Sapir 2002: 141), he also stated that "we have no right to assume that a given pattern or ritual necessarily implies a certain emotional significance or personality adjustment in its practitioners, without demonstration at the level of the individual. . . . You have to know the individual before you know what the baggage of his culture means to him" (183. Doug Hollan's seminal research has shown empirically how much this is in fact true, and the perils of an essentializing "culturalism." See, for example, Hollan 1992, 2000). An obvious corollary to this, I would add, is that a close study of individual dynamics is necessary to fully understand how the individual not only creates private meaning of shared cultural material, but negotiates his or her cultural milieu on the basis of such private meanings, thereby promoting social and cultural change. Aside from foreshadowing an epistemological position that psychological anthropologists have only recently fully reaffirmed (see, for example, LeVine 1982 [1973], Hollan 1992, Good B. 2012a, Good B. and Good M., 2017), Sapir's approach countered vehemently a classical Freudian interpretation of ethnographic fieldwork and material (see, among many others, Spiro 1965).

As will become clearer later in the book, in the field I operated with a "person-centered" ethnographic methodology (in addition to a standard participant observation approach), through a long-term series of one-on-one interview sessions with a select number of informants. Such interview sessions were carried out in a psychodynamic manner, leaving free rein to my informants to express their emotions and affects, associative thoughts, and transferential processes, while trying occasionally to elicit details and

---

[4] Importantly, conflict is one such kind of relations of interdependence. See below for details.

elaborations from them where I thought that conflicting aspects of their subjective experiences and memories were emerging. Conscious of the inherently and unavoidably "asymmetrical" character of the relationship between ethnographer and informant (as between patient and analyst; see Devereux 1969, Aron 1991, Hoffman 1994), as well as the co-constructed realm of subjective experience that a one-on-one dialectical experience obtains in two interlocutors (Orange, Atwood, and Stolorow 1997), I will not claim to have attained any "objective truth" in my conversations with my informants.[5] However, I am convinced that the intersubjective space (a "field") created by the intellectual and emotional exchange my informants experienced with me did result in the emergence of a different, and perhaps heightened, awareness of their own states of subjectivity (or "selves"), which resonated "authentically" (according to them) with their past and present life experiences ("authenticity" in a private, psychological sense. See Bion 1962, and details in chapter 4). It is this heightened self-awareness that I relied on to propose an interpretation of their unconscious dynamics and thought processes in the present, as well as with regard to their life trajectory.

Indeed, I believe that some important contributions that this book makes are to be found within the methodological and epistemological realm. Methodologically, I am proposing here a type of "clinical ethnography" (see Good B. et al. 1982, Calabrese 2013) that entails long-term interview sessions with individual informants, protracted for months, which approximate a psychoanalytic/psychotherapeutic setting.[6] At the same time, in the practice of my ethnographic/clinical encounter with my informants, and subsequently in the analysis of the material that I gathered from it, I relied epistemologically on insights from recent strands of psychoanalytic literature that have been largely overlooked by contemporary psychological anthropology, despite the fact that they constitute a promising bridge between the intrapsychic and the social/cultural. I am referring here to the insights of intersubjective analysts like Donna Orange, George Atwood, Robert Stolorow, Bernard Brandschaft, and interpersonal/relational analysts like Stephen

---

[5] Byron Good, fully understanding the "intimacy" that exists between the role of the analyst and that of the ethnographer, writes that "anthropologists do not and cannot 'know better than' the members of society with which they work. We can only discover what everybody else already knows, or *discover in collaboration* with the members of a society what is not easily knowable, what is bound up in the complicities and the 'after the facts,' what lies at the complex intersection of the psychological and the political" (Good 2012a: 32, emphasis mine).

[6] This approach has been implemented in the past by a handful of anthropologists (Kracke 1978, Obeyesekere 1981, Devereux 1951, Crapanzano 1980), but it always represented an exception to standard practices, and it remained mostly tied to a strict Freudian orthodoxy.

Mitchell, Philip Bromberg, Lewis Aron, Malcolm Slavin, and Irwin Hoffman, among others. These two "schools" of thought in psychoanalysis had their roots in the works of, respectively, Heinz Kohut and Harry Stack Sullivan (the latter alongside Karen Horney, Erich Fromm, and Clara Thompson), and both produced theoretical material that undoubtedly expresses similar preoccupations to those developed in recent anthropological theory for intersubjectivity, positionality, embodiment, and the relational construction of the self. To this, I added also my own reinterpretation of the works by British psychoanalyst Wilfred Bion and his last coworker, the Italo-Brazilian analyst Armando Ferrari (originally trained in anthropology), as well as contemporary Bionian analysts (mostly Antonino Ferro, Giuseppe Civitarese, and Riccardo Lombardi). Their subtle and elegant theory of mind is highly amenable to be integrated with cultural and social inputs.

Most of the interaction that anthropological theory has had with psychoanalytic theory over the decades has been through the foundational texts by the main figures in historical psychoanalysis, namely Sigmund Freud, Melanie Klein, and the ego-psychologists and Kleinian analysts who followed their lead. From among the other currents of mainstream psychoanalytic thought, Robert LeVine (1982) picked up elegantly on the contributions by his teacher and mentor at the Chicago Psychoanalytic Institute, Heinz Kohut (the creator of the so-called self-psychology school), and incorporated Kohut's insights into his own ethnographic and theoretical works. Other anthropologists have occasionally and loosely employed Lacanian insights to make sense of mostly broader social phenomena (see, for the most recent cases, Billé 2015, Mikkelsen 2016). By and large, however, Freud and Klein, and their successors, have held, and still hold, the high ground among anthropologists.

Freudian ego-psychologists and Kleinians, though theoretically divergent in many important respects, both maintain by and large an ontogenetic/ archeological approach to their clinical work. By this I mean that for both schools of thought and practice, the main objective of the analyst's interaction with their patients, and the major avenue through which they envision psychological recovery, is the uncovering of certain early-life disruptions, injuries, or failures in the relationship between individual and caregiver (mostly mother). Such short-circuits in the patient's early experiences would hopefully be retrieved (if repressed), relived, and expressed (through unconscious transferential processes toward the analyst), and worked through in session during the analysis, in order to attain a conscious acknowledgment of the problem at hand and foster its overcoming, including its painful psychological symptoms. The main question that this approach to psychoanalysis wants to answer is largely a "why" question: which early-life events made this patient the person he or she is today? *Why* is he or she this way today? Cultural and social worlds, interactions, and meanings that anthropologists consider crucial for the full formation of an individual and its psychological dynamics are mostly out of the attention of ego-psychology and Kleinian analysts, who are mainly engaged in the study of *intrapsychic* processes—in a way that George Stolorow has recently portrayed in these words: "One isolated mind, the analyst, is claimed to make objective observations and interpretations of another isolated mind, the patient" (Stolorow 2013: 384).

The intersubjective and relational clinicians on whose works I most rely, on the contrary, have a radically different understanding of the relation between patient and analyst, as well as the intended role of the analysis as a whole. The major question they try to answer is a "how" question: how do individuals *manage* their own predicaments at this point in life? How can one fully express such predicaments (to oneself and to the analyst)? What kinds of psychic processes and mechanisms, in the moment of the session and in the daily life of the patient, are derailing the patient's capacity to live a functional life? The focus is on the here-and-now of the session as well as on the current life of the patient, instead of the there-and-then of the early-life eventual injuries he or she might have suffered. Furthermore, this entails a more critical look at the interaction between the analyst and the patient during each session, and the way the encounter of two socially and culturally constructed subjectivities creates an intersubjective space (a "field") in which analytic work can be carried out and psychological growth achieved. On the other hand, though, and most important for an ethnographer doing anthropological work in the field, this also allows an interest for the social, cultural, and even political environment in which the patient operates (i.e., the patient's *interpersonal* context, past and present), which is legitimately considered a *locus* for the individuation of dynamics and processes that contribute to the patient's subjective formation and, possibly, pathogenesis. The existence of *intrapsychic* processes and unconscious dynamics is of course not dismissed by this approach, but rather is constantly pitted against the background of interpersonal experiences, equally formative, that the patient might go or have gone through. In the words of Robert Stolorow:

> *Patterns of intersubjective transactions within the developmental system give rise to principles (thematic patterns, meaning-structures) that unconsciously*

organize subsequent emotional and relational experiences. Such organizing principles are unconscious, not in the sense of being repressed, but in being prereflective: they ordinarily do not enter the domain of reflective self-awareness. These intersubjectively derived, prereflective organizing principles are the basic building blocks of personality development, and their totality constitutes one's character. (Stolorow 2013: 383)

It should be apparent to the eyes of an anthropologist how much this specific understanding of the psychological life of an individual is apt to be infused with those shared patterns of meaning and symbolic production (always interpersonally and intersubjectively acquired) that we call "culture." The centrality accorded by this approach to the interpersonal and intersubjective "field," as the *locus* for the shaping of the individual's subjectivity and psychic functioning, is more consistent with the theoretical developments and empirical findings of anthropological and ethnographic research than classical Freudian or Kleinian (and their followers') psychoanalytic theory is, based as it is largely on intrapsychic mechanisms and processes. And indeed, in the words of Donnel Stern:

The interpersonal field is a continuous, inevitable, social aspect of human living. It is not specifically a psychoanalytic conception in [my] frame of reference but an omnipresent, concrete, empirical reality, a sociological and psychological phenomenon that permeates and helps to constitute every moment of every human being's life. It is not possible for a person to exist outside a field. Even when one is alone, one is the product of the interpersonal fields in which one has come to be, and one's experience continues to take its meaning from them, and to contribute to their continuing evolution. (Stern D. B. 2015: 389)

Stern's words echo nicely those by anthropologist Suad Joseph (Joseph 1999), who two decades ago posited that deep interpersonal connectedness and the mutual blurring of self-boundaries, far from being the sign of a pathological fragmentation of the bounded ego (as many ego-psychologists would claim. See, for instance, Hartmann 1958), are in fact a constitutive aspect of *healthy* subjective experiences in certain specific cultural contexts.

Additionally, in my quest for a non-dogmatic, comparative, and empirically based set of clinical insights to a better "reading" of the subjectivities that my informants displayed in session with me and in their daily life environment, I found extremely useful the theory of thought that Wilfred Bion and contemporary Bionian analysts developed over the decades, and are still very much expanding. Indeed, Bionian clinicians have certainly established a "field theory" of their own, which shares some of the same principles as that of intersubjective and interpersonal/relational analysts (see, for example, Ferro 2002, 2009, Ferro and Basile 2009, Civitarese 2010, 2013, Chianese 2007), and yet their minute attention to the workings of the mind and its thought processes, as well as their investigations into the very emergence of meaningful thinking activity in the individual, is something that sets them apart from other psychoanalytic schools of thought, and that I found very helpful in dealing with my informants' inner lives.

By making use of the interpersonal and intersubjective "field" that originated from interaction between me and my informants in session, I detail the formation of, and dynamic relationship between, conflicting and alternative subjectivities in my informants, which pertain to their current life as well as to their past vicissitudes. The field that obtains in session is "a space in which incompatible selves, each awake to its own 'truth,' can 'dream' the reality of the other without risk to its own integrity . . . the reciprocal process of active involvement with the states of mind of 'the other,' allows a patient's here-and-now perception of self to share consciousness with the experiences of incompatible self-narratives that were formerly dissociated" (Bromberg 1996: 278). The "dreaming" that Bromberg is referring to should not be interpreted as the conventional dreaming activity that happens during sleep-time. Instead, it is akin to what Antonino Ferro, elaborating on Wilfred Bion's similar idea, calls "waking dream thoughts" (Ferro 2009): the transformation of unsymbolized and unformulated sensuous impressions, protoemotional experience, into symbols, and hence thought. Dreaming, in this sense, becomes then not the exclusive pertinence of sleep-time activity, but the everyday process of bringing to awareness and consciousness previously "unformulated experience" (Stern D. B. 2010), emotional cacophonic debris (the "beta-elements," in Wilfred Bion's terminology). From this standpoint, "we are always dreaming, even while awake" (Stern D. B. 2013: 633). I will try to show in this book that the experience of two minds interfacing each other (mine and my informant's) created the intersubjective space in which it became possible for my informants to "dream" and symbolize incompatible subjective states and self-narratives (1) that they had construed for themselves in various conjunctures of their lives, (2) with which they were still living, and (3) between whose spaces they managed to stand, without losing their own sense of overall integrity and coherence.

To sum up, the psychoanalysis that I conjured to help me understand my informants as *both psychological and cultural beings* is "no longer [a] psychoanalysis that aims to remove the veil of repression or to integrate splittings, but [a] psychoanalysis interested in the development of the tools that allow the development and the creation of thought, that is the mental apparatus for dreaming, feeling and thinking" (Ferro 2006b: 990). At the same time, and in the same vein, the psychoanalysis that I envision here

> *[G]ains its best opportunities not from focusing primarily on the influence of the inner world on the outer one, and certainly not from a focus on the external one, but on* the continuous interaction of these two kinds of meaning, and their mutual constitution of one another. *Our grasp of the external world, of course, is deeply influenced and informed by the internal one. But the internal world is also shaped by the events in the world outside it, and especially by trauma . . . in practice,* the two worlds are completely entwined with one another. (Stern D. B. 2013: 638, emphasis mine)

On my part, I *integrate* the insights and cues provided by all these clinicians with the cues stemming the cultural, social, and political landscape wherein my informants lived. In other words, through ethnographic practice I position psychodynamic processes within the behavioral environment (Hallowell 1955) in which my informants operated. This I believe was the kind of integration that Edward Sapir envisioned during the years in which he collaborated closely with Harry Stack Sullivan and his academic journal *Psychiatry* in the 1930s. And it is likely also in this vein that Irving Hallowell urged for a *psychodynamic* register of investigation geared toward a fuller understanding of social and cultural phenomena. After all, as he stated, "it is quite true, of course, that the self known to the individual may not represent a 'true' picture" (Hallowell 1955: 80). Indeed, more recently, Byron Good has suggested the need for a hauntological perspective on subjectivity, which should take advantage of psychoanalytic and psychodynamic theories to "[reflect] on that which is hidden—that to which one does not have access to without concentrated reflection, or many times even without such reflection—some of which is hidden consciously and purposely for political purposes, but much of which is rooted in fears and desires, far beyond any rational control" (Good B. 2012a: 27).

While I am certainly critical, from an anthropological standpoint, of the insufficient attention that *all* schools of psychoanalytic thought devote to the

sociocultural environment in which the individual operates, I am still convinced of the validity of their overall approach to the study of the human
mind and behavior, and persuaded that such usefulness would be even
enhanced if interwoven with the web of meanings and subjectivity-shaping
processes we find in culture, social arrangements, and political dynamics.

An intellectual genealogy

In a recent article, anthropologist and psychoanalyst Douglas Hollan asks
a question that becomes extremely relevant to all I have said so far: "How
do we as anthropologists pick and choose among the very diverse psychoanalytic concepts available to us, especially when these concepts may come
embedded in very different, and sometimes contradictory, theoretical
assumptions about human behavior?" (Hollan 2016: 508). I believe Hollan's
is a crucial question for whoever works in such proximity to psychoanalysis
as I have done, and for the contextualization of the intellectual genealogy of
this book, and its author.

So, how did *I* "pick and choose" the psychoanalytic theories and thinkers
that I then used in my research? For one, since the early stages of my research
it became quickly clear that no single one, well-structured, and coherent
theoretical system would really be so all-encompassing and far-reaching to
cover alone and successfully all the possible variations of human psychological dynamics. I thus took advantage of multiple positions and insights from
several clinicians, not always in the same theoretical stream, in a comparative
and integrated manner.

Additionally, however, there is a more personal, and less "detached," side
to this story as well (as is often the case). I was in psychoanalytic therapy
as a patient for many years, starting in my late teens. When I came to the
United States in 2007 from Italy, my home country, at age thirty-six, I used
to joke with the clinicians whom I got to know at the Massachusetts Institute
for Psychoanalysis (MIP) in Boston that I, like themselves, had some professional qualifications and clinical competence too, because, after all, I had
become by then a "professional analysand"! The joke was not totally a joke,
though. My experience with psychoanalysis has had a profound mutational,
enabling, and, overall, positive effect on my life. It was certainly at times
painful and hard to bear, like the psychological reality that brought me to
it in the first place, but it allowed me to gain a much broader and deeper

knowledge of myself, my inner functioning, while at the same time it allowed me to acquire those tools necessary to operate on myself autonomously, and *manage* the "glitches" that made my existence problematic. Over the years I came to appreciate and understand the centrality of certain characteristics of one's own inner life—unconscious processes—and how much they are "responsible" for a good deal of one's outward behaviors and expressions. This is also the main reason, I believe, why I became so interested in the same kinds of processes in my own informants when I started doing my fieldwork.

The major part of my "career" as an analysand I spent with an Italian analyst who trained at the psychoanalytic institute founded in Italy by Armando Ferrari, an Italo-Brazilian analyst who was a student and then close colleague of Wilfred Bion in his final years (the second half of the 1970s). While Ferrari's work and theoretical contributions owe very much to Bion's, he later became fully his own clinician and thinker (just as Bion did with Melanie Klein, his early analyst and mentor). Though I certainly did read Bion extensively, his ideas that I incorporate in my writing come to me also through the lens of my analyst's and Armando Ferrari's theorization (see, for instance, Ferrari 2004). It is something that I obviously came to know very intimately, whose strengths and weaknesses I could pit against my own personal experience, and which I felt confident to use carefully in the relationship with my informants.

Yet, in all fairness to myself, I was not only a "professional analysand" after all. Before going to the field for my long-term fieldwork (2012–2013), I received two years of training in clinical psychoanalysis at MIP in Boston. MIP is a very inclusive, comparative training institute, which strives to avoid dogmatisms and offers its trainees knowledge in the diverse "souls" of psychoanalytic theory and practice. Among the many authors whose works we had to engage at MIP, were the interpersonalists/relational and the intersubjectivists. Though I had known this literature for many years, MIP provided the chance to dig deeper into, and discuss openly, the positions and experiences presented by these clinicians. While I was in training, I was also continuing to visit my fieldsite on a regular basis for shorter periods of time, which gave me the opportunity to tentatively look at some of my informants' narratives and their relationships with me through the epistemological lens provided by some of the literature that struck me the most. As I said, I had already by then developed the sense that a clinical approach that paid due attention and consideration to sociocultural factors in each individual's life might be more in tune with not only my anthropological sensitivity, but also with what I was in fact observing on the ground. Indeed, if there is a slight "advantage," so to speak, that the clinical ethnographer has on the clinical analyst is that the former, unlike the latter, has the privilege of being able to follow his or her informants through their daily vicissitudes, and compare the dynamics of a one-on-one clinical session with the contradictions possibly emerging in the informant's public life. In this regard, the insights I gained from the works of interpersonal/relational and intersubjective analysts proved to be extremely useful in my field research.

In the end, for me, choosing to embark in the study of the social and cultural dynamics entailed by my fieldsite through the prism of the inner lives and psychic processes of my informants, was profoundly motivated by my own existential trajectory, and the way psychoanalytic practice (as analysand, who shares this practice on equal terms with the analyst) marked my life and my understanding of human phenomena. I came to realize that the exercise of psychoanalysis, and psychotherapy more in general, shares deep ties with ethnographic research. It has the long-term goal of attaining psychological growth in the patient (a healing?), but it does so by fostering the "understanding" of one's own (the patient's and the analyst's) paradigms of psychic functioning and interpersonal models of engagement.[7] While the first objective (healing) may not be within the reach, or even in the intention, of the clinical ethnographer, the second (understanding) certainly is, and I believe may yield results in ethnographic research.[8]

Engaging masculinity

Feminist scholars Anne Sisson Runyan and V. Spike Peterson write that:

> *The power of gender operates as a meta-lens that orders and constrains thinking and thus social reality and action, thereby serving as a major impediment to addressing inequalities and the global crises we begin to explore below that stem from, sustain and even worsen inequalities. . . . At its deepest*

---

[7] I am referring here not only to an "intellectual" knowing, but most of all to what Wilfred Bion terms cryptically as the "O" in a relationship—those rare moments of reaching a mutual unconscious understanding through the intersubjective operation of two minds. See Bion 1967b.

[8] In fact, Robert LeVine, forty years ago, spelled out an iron rule whose validity holds still strong: "Individuals, and only individuals, can be psychoanalyzed. Customs, institutions and organizations cannot be, and any attempt to do so involves dispensing with those elements in the clinical method that give psychoanalytic assessments their validity" (LeVine 1982 [1973]: 209).

level, the power of gender as a meta-lens continually normalizes—and hence depoliticizes—essentialized stereotypes, dichotomized categories, and hierarchical arrangements. In these multiple and overlapping ways, the power of gender is political: it operates pervasively to produce and sustain unequal power relations. (Runyan and Peterson 2013: 8)

This passage contains one of the most powerful and incisive assessments of the way gender ideologies impact and direct deeply engrained social, cultural, and political dynamics. Furthermore, it offers a convincing roadmap for an increased sociological and anthropological scrutiny onto the constraining nature of gender upon human interaction. Indeed, ethnographic literature and anthropological theory have produced a vast amount of work in this realm, especially during the decades of feminist and post-modernist scholarship (from the mid-1970s through the 1990s: see particularly Leacock 1981, Sanday 1981, Ortner 1996, Rosaldo and Lamphere 1974, Lamphere, Ragone, and Zavella 1997). To remain with Runyan's and Peterson's "meta-lens" image, I will say that in this book I work *from within* the meta-lens of the "power of gender," though fully acknowledging its existence and operation. I will leave the main social ramifications of the operations of the meta-lens somewhat in the background. It was not my intention to participate in the debate about perceived structures for male domination, hegemonic patriarchy, and the consequent denial for women of a conspicuous and meaningful participation in the public and political sphere of their own communities (all of which irrefutably apply also to Pashtun society, as we will see from my informants' accounts). Masculinity in this book represents the cultural prism through which I explored the subjectivity of my informants. In this sense, this book is not about masculinity *stricto sensu*. I made instrumental use of this cultural idiom, which I knew was central to the lives of my Pashtun male informants, in order to investigate their psychological dynamics.[9]

---

[9] Of course, some observers will see the psychic struggle that my informants go through as a direct outcome of being betwixt and between "modernity" and an otherwise "traditional" world. It may be a legitimate, and still not exhaustive, way to look at the material I present, only so long as we realize and acknowledge that modernity, for most Afghan people since the early 1980s, has been represented de facto by a continuous state of conflict largely instigated and supported by outside actors, with strategic and political global interests that vastly outsized Afghanistan itself, of which the country was simply the warring ground. The disruption and change that war brings along is certainly at the center of the suffering that I am investigating in this book and may be singled out as one of the main roots of it. So, yes, in these terms, being caught in between "modernity" and "tradition" is one of the independent variables constitutive of the phenomena I study here.

Let me expand briefly on this point. In this book I analyze a stage in my informants' efforts at negotiating, restructuring, and even "manipulating" their received cultural narrative about masculinity, that rests in the background of their public performance of it, and on which such performance depends—that is, their often unconscious psychic and even cognitive dynamics. As I will detail in the following chapters, cultural idioms for an appropriate masculinity in a Pashtun social environment are well-defined, rather strict, and enjoin strong conformity in public behavior. This does not mean, of course, that all Pashtun men display the same set of outward behaviors, or that those behaviors have the same meaning for individual Pashtun men. It does mean, however, that those who run astray of the social expectations for appropriate masculine behavior often pay a dire price in social rebuttal and reproach, which negatively affects their and their family's claims to respectability and honorability. In my work I investigate what my informants "do" psychologically with such idioms of masculinity they are supposed to share with their peers, how this inner "doing" shapes their outward behavioral expressions, and, consciously or unconsciously, operates on their awareness of the "power of gender." Of course, I do discuss the performance of masculinity they enact in their social context, for I had the opportunity to follow many of them in their daily lives and compare what came out of our intimate and lengthy conversations with what they displayed publicly. Yet my primary focus was on the psychic dynamics these men were forced to develop in order to cope with what society expected them to be, which often resulted in a significant degree of inner conflict, ambivalence, and, ultimately, psychological suffering. It is mainly for these reasons that I should "warn" the reader that this book is not strictly about masculinity.

Additionally, I will say that gender roles, and hence masculinity as well, are widely understood today within the social sciences also through the analytical lens of "performativity," which Judith Butler theorized in a set of writings, mainly during the 1990s (Butler 1988, 1990, 1993, 1997). It is not that the subject "does" or "performs" gender, as it were, in Butler's view. Rather, performativity is *constitutive* of gender, and of the gendered subject (Butler 1990: 25). To be fair, Butler's language in this regard is confusing, and her vacillations between a subject that is aprioristically pervaded by and subordinated to the workings of "power" and "performativity," and a subject that is allowed some degree of agency (and thus itself "does" gender), have been rightly pointed out and criticized by later scholarship (see Brickell 2005). Yet the performative aspect of gender norms has been picked up by most of subsequent anthropological scholarship, while it was even somehow anticipated by some previous authors as well (Peristiany 1966, Herzfeld 1985, Gilmore 1987, 1990, Gutmann 2007 [1996], Mahmood 2005, Ghannam 2013). Whereas the social *consequences* of the performance of gender roles (in their *performative* aspect, as a condition for the gendered formation of the subject) have been a widely accepted basis for the understanding and interpretation of what gender, and in my case, masculinity, means, on the other hand I have focused in this book on the psychic *antecedents* of such performance. I did so upon the conviction that in fact the subject *does* participate agentically in the formation of that set of symbolic, shared meanings that we call cultural idioms of masculinity (and in so doing, it is not only performatively gendered). My position is a decidedly *ethnographic* one, as opposed to the mostly *ontological* position developed by Butler. By this I mean that I responded to the experiential vicissitudes that my informants expressed to me and had to cope with, while refraining from observing them through the lens of an ontological category—the "subject"—devised in abstraction from real individuals and endowed with properties that do not belong necessarily to any of them (a deductive, instead of inductive, process). My informants faced a set of norms and rules about culturally appropriate masculinity into which they were enculturated. These norms were thrown at them (metaphorically), and they had, one way or another, to react and adjust to them. Alongside my informants, I investigated how they negotiated and reworked psychologically these norms. From these negotiations (often painful, ridden with anxiety and contradictory) emerged the public behavior they displayed as men and which I, to a certain extent, was able to witness and pit against what I thought motivated it within their subjectivity.

Matthew Gutmann writes that masculinity is "anything that men think and do," and "anything men think and do to be men" (Gutmann 2007 [1996]: 386). But how does that come to be? What prompts men to say or do what they say or do about masculinity? Furthermore, is writing about men different from writing about masculinity? It might seem in certain instances that this could be the case. Yet, if the subject is always, inescapably gendered, in all its social and cultural expressions, can we really conceive of a discourse about "men" that is not also, inherently if not overtly, a discourse on masculinity?

This also calls into question the narrative style that this book embraces. In each chapter, the reader will find significant excerpts from my long-term interview sessions with my informants, which I analyze and link together

in the effort to describe how my informants came to make certain behavioral choices in their lives. Narrative, or "verbalized thought," as Wilfred
Bion termed it (Bion 1967a), is of course the main medium through which
practices such as psychotherapy and psychoanalysis proceed in their investigation of human subjectivity, in its conscious and unconscious expressions
(though, for a sophisticated appreciation of nonverbal and non-symbolic
meaning-formation and therapeutic action, based on the interpersonal and
intersubjective encounter, see a recent strand of psychoanalytic thought
exemplified in BGPSG 1998, 2007).[10] Yet also in recent anthropological
theory, the centrality of personal narratives has found its due recognition
toward the understanding of the individual's inner processes, beyond their
phenomenological aspects (see Good B. 2012a). In this regard, I find myself in complete agreement with Janis Jenkins when she writes that "the
analysis of narrative is a methodological locus for understanding subjective
experience and transformation of self in a cultural context through stories or
descriptions of conditions and events" (Jenkins 2015: 9). Byron Good goes
even deeper, suggesting that "in anthropological research we need constantly
to *listen* to what is unspoken, unsaid, repressed, unspeakable—in politics and
in everyday life—and to attend to our own resistances to knowing as much
as to the complex forms of resistance to knowing of those with whom we
work" (Good B. 2012a: 31, emphasis mine). In the specific realm of idioms
of Pashtun masculinity, the personal narratives that I analyze in the book
give me the opportunity to explore masculinity as analytical category, and
not only "men" as a social category, precisely because they give shape to those
private processes, both conscious and unconscious, that ultimately make the
performance of masculinity possible. If we all can agree with Judith Butler
(1988) when she writes that it is the repetition of certain behaviors premised
on cultural norms that makes the concept of gender "alive" and hence visible, and yet that such norms get somewhat modified with each iteration (for

---

[10] In a recent article (BCPSG 2007: 3), the Boston Change Process Study Group elaborates on the
presence and impact of what they term "implicit relational knowing" that supposedly emerges early
in pre-verbal infants: "Implicit relational knowing is thus a form of representation. In using the word
'knowing' we do not imply a symbolic process. It is the intuitive sense, based on one's history, of
how to be with another. It concerns knowledge and representation that are not language based, so
that studies of pre-verbal infants provide an unencumbered field for its study. In brief, implicit relational knowing is based in affect and action, rather than in word and symbol. It is also unconscious
but not under repression. Accordingly, it can be brought to consciousness and verbalized, but usually with much difficulty." This view carries implications also for analytical practice, and its reliance
on verbal "interpretations" (see BCPSG 1998). The BCPSG was historically composed by Daniel
Stern, Louis Sander, Jeremy Nahum, Alexandra Harrison, Karlen Lyons-Ruth, Alec Morgan, Nadia
Bruschweilerstern, and Edward Tronick.

there is never full duplication in repetition), then it becomes apparent that what I analyze in these pages are the personal, inner processes that allow repetition to happen in the public behavior of my informants (though idiosyncratically subjectivized by each individual), as well as the deviations from those very norms.

In aiming to explore cultural configurations of Pashtun masculinity less from a political and sociological perspective, than from a truly experiential and individual-centered one, my approach runs more along the lines of the ethnographic research that Marcia Inhorn has recently pursued (Inhorn 2012), which focused on the problem of male infertility and artificial insemination among Arab Middle Eastern couples. Inhorn used the difficult medical condition of her male informants to provide insight into her informants' sense of manly worth. The portrait that comes out of her research sheds light on today's position of cultural idioms of masculinity in the Arab Middle Eastern context, as well as on the struggle that men are putting up in order to adjust to an objectively punishing condition, while at the same time operating on their society's cultural idioms through their acts of "resistance" and assertion of personal (and masculine) agency. Likewise, my research focuses on the lived experience of my male informants, within a cultural framework of demanding and taxing expectations about appropriate masculine behavior, onto which they, as we will see, operated either as reinforcing agents, or disavowing ones. I no doubt rely (as did Inhorn) on the scholarship that sociologist Robert W. Connell produced on the idea of hegemonic and subordinate masculinity (Connell 1987, 1995). Connell too, as I did, worked *from within* the meta-lens constituted by the power of gender, and concerned himself specifically with the relations of power and subordination *within* the male sex, rather than *between* the male and female sexes. He emphasized that, just as it is undeniable that there exists a social relationship of unequal power between men and women in most societies in the world, it is equally undeniable that there exists a hierarchical stratification of masculine roles within male-only milieus as well, with a hegemonic role on top, and many others subordinated to it, in relation to the sociocultural context. More recently, Demetriakis Demetriou refined Connell's concept, pointing out that also within a hegemonic category of masculine cultural idioms, we should talk of *masculinities*, mutually competing, instead of simply one hegemonic masculinity (Demetriou 2001). The analysis that Connell and Demetriou carry out is particularly fitting for a strongly homosocial context, such as the Pashtuns'. We will see from my informants' accounts of their daily lives' vicissitudes and trajectories, that they had constantly to grapple with, and navigate through, a cultural world that provided them at times with the "luxury" of being able to choose (if often unconsciously) from among several hegemonic masculine roles, in competition with each other, while at other times it forced them to embody one only configuration for a hegemonic masculinity (or else resign to suffer the social consequences of being the bearers of a subordinate one). Relations of power, expressed often through aggressive and even violent behaviors, were a central component of my informants' lived experiences among their peers.

There was also a more personal, subjective side to my choice of "masculinity" as the "turf" on which to play out my research. Manliness, its cultural significance, its ways of social expression, and its public role in society, held a particular spot in my personal experiences. After graduating from college, in 1997, I served three and half years in the Italian army, prior to moving to freelancing journalism (as a war correspondent in the Middle East). I returned to my military unit in 2004, and served there three more years, after acknowledging that working as a freelance journalist was not enough to pay the bills. As an officer in an army's regiment, I lived and operated within a very distinct subculture, encapsulated by the "regular" Italian sociocultural environment. In my military world, moral values such as honor, respect, sacrifice, and ethical qualities such as courage, valor, and fearlessness, were integral parts of the role that each of us played qua soldiers. Our manliness, our manly worth, depended on how well we could demonstrate we possessed such moral values and ethical qualities. Furthermore, my unit was a paratroopers' special unit, trained for particular tasks, which magnified and enhanced in its members the necessity to display these ethico-moral requirements. Only such a man could be considered an adequate and dignified soldier, and, more important, deserving to belong to our special unit. Note that, at that time, women were not yet allowed to serve in the Italian army, as they are now. To this very day, however, women are not allowed to serve in my former unit. Sure enough, there were those among us who took these ideal standards more seriously than others, and others who more pragmatically adjusted to a reality that did not always live up to those standards. Yet the ideal model was clear to everybody, and everyone was implicitly evaluated against this yardstick. During my last three years of military service, the period 2004–2007, I started to seriously deconstruct the meaning that these cultural patterns held for me, and to radically question the very essence of the role that I was embodying as a soldier (especially during my deployment to Iraq, in 2006). Yet the features of what had constituted a crucial part of my experience in uniform (manliness, and masculinity as a whole), remained inescapably a focus of interest for me.

We will see in this book that narratives regarding honor, respect, courage, and fearlessness are central to the framing of a *culturally* appropriate and adequate masculine subjectivity among Pashtun men. When I realized, through my studies and my preliminary research visits, that Pashtun men's cultural and social universe revolved around patterns that seemed so familiar to me (though enacted in very different ways, of course), I felt that it would "make sense" for me to follow this research lead as a cultural anthropologist. In the end, through the exploration I carried out in my research about Pashtun men's subjectivity, sense of masculinity, and manly worth, I explored mine as well.

The methodological framework

Over the first years of my research in Afghanistan, from 2009 until 2012, I spent relatively short periods of time in the country on a regular basis: three months during summers, and three weeks during the academic Christmas breaks. These repeated visits gave me the opportunity to refine the objectives of my research, get acquainted with the social and cultural reality of my fieldsite, and become proficient in Pashto (the local language). More important, though, they allowed me to build a network of friends, acquaintances, and occasional contacts, who constituted a reliable and trustworthy environment, on the basis of which I could safely conduct the long-term fieldwork that was to come. Indeed, by summer 2012, when I started my ten months' long period of continuous research, the region of my fieldsite had become a hotbed for the insurrection against the US-led international military coalition and the Western-backed Afghan government that had sprung up after the demise of the Taliban regime in late 2001. The security situation in Nangarhar province was rapidly deteriorating by the week. Without the people that I had come to know well during the previous visits, it would have been impossible for me to reside stably in my fieldsite for so long. Through one of them I was able to rent a house for myself in Jalalabad, the capital of Nangarhar province, in the southeast of the country, near the border with Pakistan.

In this house I held many of the interview sessions on which my person-centered research was based.[11] I would sit in one of the rooms with my informant, and with nobody else present. The series of sessions ranged from a minimum of five, to a maximum of thirty-seven (with Baryalay). Each of them lasted between forty-five minutes and two hours, depending on the circumstances. In this book, I will present the detailed account of the series of sessions I had with four informants, and a shorter profile of four more. I met only one of my informants exclusively at my place. The others I had known for several years, and had participated with them in the social activities and events they attended either in Jalalabad, or in their home village, or even in Kabul. My informants came from diverse backgrounds: urban, rural, educated, illiterate, semi-literate, religiously minded, secular (relatively to the context), even iconoclastic. Some had been refugees in Pakistan, others had never left Afghanistan, even at the height of the conflicts. Some spoke some English, others good English, others only Pashto. The age range is also very diverse: one of my informants was sixteen when I first met him, and (barely) twenty-one when I left. His grandfather, also an informant, is in his mid-seventies. Some of my informants lived in Jalalabad, others in rural villages, a few more in Kabul, and in one rural district of Paktia province. I met them in both Jalalabad and their home villages (or Kabul). From my permanent base in the house I rented in Jalalabad, my life in the field was spent moving from one place to another, in order to find the best logistical solution to meet the people I wanted to talk to, in the least possible "disruptive" way (socially, culturally, and with regards to security). In this sense, my fieldsite proper comprises a large area of Nangarhar province, given that many of the people I worked with lived in rural villages scattered around several districts of the province. From this standpoint, it amounted to a "multi-sited" fieldwork. In fact, security concerns represented the major variable upon which my meetings with informants were arranged, particularly during the last months of my fieldwork. With each informant I met, we had to consider the security situation of the area where we decided to meet, at any given time during my research. We took into consideration the most recent range of activities of the insurgency, and what was expected from them at that time. Sometimes Jalalabad was the only safe place where I could meet one of my informants, while at other times it would be viable to meet in his village, and maybe stay

---

[11] The main theoretical guidelines of this type of ethnographic work were set by Robert LeVine, in *Culture, Behavior, and Personality* (1982 [1973]), and were spelled out more pragmatically in later papers by Robert Levy and Douglas Hollan (Levy and Hollan 1998, Hollan 2001).

there for a few nights. Sometimes one of my informants would not consider it safe for him to be seen in Jalalabad at all, and at the same time it would not be safe for me to meet him in his village. In these cases, we would meet in a third location, usually a friend's house, away from other people's gaze.

I must say at this point that the safety of my informants was obviously my very first preoccupation, and I took the matter very seriously. My most important concern was about what I *should not* do, given the potential negative consequences, much more than about what I *could* do. Yet, even more than my attention to the context and situation, what was critical was the attitude of my informants themselves. We will see later that one important cultural value in Pashtun society, which many hold dear as a reflection of the family's respectability and honorability, is that of hospitality (*melmastia*). Yet I quickly realized that the norms, rules, and moral values undergirding melmastia were very often operationalized in a strictly pragmatic way. What I mean is that the obligation to show hospitality was strictly conditional, not absolute, and remained truly an obligation only so long as the lineage or household in question could maintain a modicum of safety. I was very candidly told that in an ecological context of such great violence, volatility and uncertainty, nobody was ready and willing to seriously risk their own and their family members' lives only to ensure that some cultural custom be faithfully respected. Nobody would expect that of them, I was made to understand, even within their own local circle. All the more when dealing with me, a stranger and outsider (from multiple standpoints). This was a useful lesson on how culture and cultural practices are certainly central leads in any given society's dynamics, but should not be interpreted as iron-clad, inelastic, ineffable behavioral ruts, out of which it is impossible to extricate oneself.

The people I worked with were extremely keen observers and were cognizant of what was happening in their surroundings. They knew exactly how far they could go before they would suffer some sort of serious repercussion. Furthermore, in contexts such as these, where your enemy is never a total stranger, they would monitor continuously the pulse of the situation step by step. Nothing would have happened to them without a series of discussions, warnings, and the like, with people of their long-standing knowledge. The candor and frankness with which all these subtle details of social and political coexistence were explained and shown to me made me confident that, following their lead on what I could and could not do and ask of them, all of us would be safe.

The setting, internal features, and rhythm of the sessions I had with my informants tended to mimic a psychotherapeutic or psychoanalytic session. With some of them this aspect was more marked, with others less so (we will see how in more detail during the discussion of the material). The reader will have some idea of how our sessions proceeded through the verbatim excerpts of them that I present in each chapter. So, while I aimed at the emergence of thought processes, subjective experiences, and unconscious dynamics in my informants—which has its equivalent in the psychotherapeutic/psychoanalytic session—I will certainly not suggest here a complete identity between the latter and our interview sessions. There are definitely important, distinctive differences between the two experiences. The most obvious one is that it was *I* who went to them, who "needed" them, and not vice versa, as it happens in a patient-analyst dyad. Also, I did not *intend* to provoke any psychological change, development, or growth in my informants. Likewise, my informants did not feel (at least openly) that they had any problem they needed help with from me. In short, my role in relation to them was not intended to be a "therapeutic" one—though this does not mean that a therapeutic outcome might not have been attained anyway. Related to this point is the issue of the temporal length of our relationships. I spent approximately fifty-five full hours of face-to-face conversation with Baryalay, the informant with whom I had the highest number of sessions (which averaged one and half hour each). Each analytic session usually lasts for forty-five minutes. In "analytic time," I had sixty-eight sessions with Baryalay. Considering an average of three sessions per week in the course of an analytic treatment, our sixty-eight sessions would have spanned twenty-two weeks, which equal about five months and half. Five and half months are considered today as the starting period of a full-fledged analytic treatment, a period of time that cannot possibly warrant the resolution of profound psychic conflicts and suffering.[12] It is a period in which usually the analyst manages to gain a degree of insight into the system of psychic and mental functioning that the patient is using in order to navigate his or her own life and personal vicissitudes. The patient's main problems come to the surface (though sometimes it takes much longer than that), and the analyst is, on average, able to become aware of the mechanisms

[12] It is interesting to note that in Freud's intentions, a psychoanalytic treatment should have lasted about six months only, beyond which it should be brought to a conclusion. This emerges from a close reading of his paper "Analysis Terminable and Interminable" (Freud 1937). Psychoanalytic theory and practice have evolved enormously since the publication of Freud's paper, and today six months is considered to be certainly an inadequate amount of time for a full understanding (and possibly "resolution," in whatever fashion it may happen) of a patient's deeper conflicts.

that enhance, or hinder, the patient's ability to cope with them. Yet it is only the beginning. My relationship with my informants could not aspire to reach further than this initial stage in understanding the complexities of their inner lives and reality. Nevertheless, during the sessions I had with each of them a rich psychodynamic work could be constructed and developed.

The contiguity between ethnographic and psychoanalytic practice

A well-known aspect of the psychodynamic exchange between two individuals, transference is conceived in psychoanalytic theory as the repetition of certain patterns of behavior and/or emotional engagement with the analyst, which the patient had with some significant other person during his or her life. In other words, the patient unconsciously addresses (and deals with) the analyst as he or she might have done with some important figures in their life, say, the mother or father. Within the analytic setting, together with the analyst, the patient will in theory be able to become aware of those otherwise unconscious patterns of behavior and emotional expressions, and "work" on them. Countertransference is represented by the conscious and unconscious reactions of the analyst (whether in the realm of fantasy, rational thought, or unconscious dynamics) to the material presented to him or her by the patient. Robert Stolorow addresses both transference and countertransference this way:

> The patient's transference experience is co-constituted by the patient's pre-reflective organizing principles and whatever is coming from the analyst that is lending itself to being organized by them. The same statement can be made about the analyst's transference [i.e., the countertransference]. The psychological field formed by the interplay of the patient's transference and the analyst's transference is an example of what we call an intersubjective system. Psychoanalysis is a dialogical method for bringing this pre-reflective organizing activity into reflective self-awareness. (Stolorow 2013: 383)

Both transference and countertransference need not happen exclusively during an analytic session, but rather are understood to be one of the ways in which the human mind works during any sort of interpersonal exchange. What changes in the analytic session is the specific attention and "monitoring" that analyst and patient, together, lend to these processes, rendering them therapeutically meaningful. Interpreting the occurrence of a transference phenomenon is precisely that—a subjective, interpretive endeavor. As such, it is perforce prone to be questioned as to its appropriateness and "legitimacy."

The case of countertransference is certainly more open to a transparent elaboration (Wilfred Bion calls this process "reverie"), because the analyst (or, in my case, the ethnographer) is not only the person who is subject to the process, but also the one who will write and elaborate about it (see Ogden [1994], for a brief but comprehensive overview on the development of the concept of countertransference in psychoanalytic theory). I have underscored in the text those instances in which I managed to become aware of the emotional reactions, and streams of thought, that my informants' words triggered in me, and the cases in which I believe the countertransferential material helped me better understand the dynamic of the exchange between me and my informants. As always happens, there have surely been instances in which my emotional reactions remained unconscious, and as such not perceived openly by me. In general, however, whatever the features and character of the countertransferential material, I am convinced with anthropologist and psychoanalyst George Devereux (as well as many contemporary psychoanalysts) that countertransference, or reverie, is a crucial and positive aspect of any close interpersonal encounter (in this respect, Devereux was much ahead of his time even by the standards of psychoanalytic theory. Tellingly, however, Devereux's intuitions are not credited in Thomas Ogden's overview of countertransference I mentioned above). Devereux, in *From Anxiety to Method* (1967), spelled out convincingly not only the reasons why, in his opinion, the work of the ethnographer is intimately related to the work of an analyst with his or her patients, but also the reasons why ethnographers need to factor their own emotions into their account of the reality they study—emotions that derive directly from the "trauma" of the protracted exposure to a subjectively unconventional and alien environment. Devereux writes that the ethnographer who declares that he or she remained a detached and objective observer, who claims to suffer no emotional backlash from the things he or she saw and experienced personally, is simply in a state of denial (Devereux 1967: 83–102 et passim). Such a state obfuscates and distorts the account itself that the ethnographer will later produce in writing. It is necessary, in Devereux's view, to accept the reality that we, as ethnographers, are the "victims" of (often negative) affects while in the field, and we must to try to deconstruct and understand them before putting hand to pen and paper.

When I definitively came back from the field, in summer 2013, I felt bitter. After four years of research, I felt resentful toward the people among whom I worked, for all the hostility and rejection that they had expressed against me. I obviously felt all the more grateful to those with whom I did manage to work closely. Yet, as will become clear through the chapters of this book, in my fieldsite I lived surrounded by hostility, acrimony, and fear. Those who decided to work closely with me were among the minority who did not see in me either a dangerous enemy, or a despicable infidel, or both, for religious and political reasons. I was often considered a potential spy, or a potential missionary, or the representative of those who had inflicted onto the Pashtun population of Afghanistan death and sorrow (i.e., the international troops), and for this reason morally responsible. Those who worked intimately with me had often to respond to public criticism, or moral reprobation from their peers. Some decided to interrupt the relationship they were having with me, while others tried to keep it as low-key as possible. Still others chose to go against the stream, and collaborated closely with me.[13]

As a result, I felt frustrated and bitter, not only because of the kind of experience that I personally had, and for the limitations that such state of facts exerted on my fieldwork, but mostly for all the personal discomfort, uneasiness, and even social problems that my friends and informants had sometimes to face because of their choice to help my research.[14] I slowly became aware of this emotional process that was taking place in me, and, when I finally realized its full extent, I took it seriously, and started to judge all my

---

[13] Make no mistake, though, those who maintained these prejudices against me were not victims of a sort of "collective paranoia": Afghanistan has a long history of very real foreign military and political interference, aggressive intelligence gathering, and covert religious proselytizing activities (there were more than a few Christian missionaries in disguise in Jalalabad while I was living there, of whom I met a couple).

[14] It is equally true, however, that over time I came to realize that my difficulties during fieldwork, which I just mentioned, sprang from the same dynamics and circumstances that were making my research possible in the first place. By this I mean that I was able to witness and become privy to the emotional hardships, psychic conflicts, and cultural contradictions that my informants were experiencing, because (yet not exclusively) of those same wartime sociocultural disruptions that Pashtun society had undergone for over three decades, and which were rendering my fieldwork methodologically and emotionally challenging. In this regard, I was reminded of the vibrant and empathetic assessment that Erik Erikson gave of his research (albeit short-term) among Native American Oglala Sioux and Yurok individuals (Erikson 1950: 98–131 and 141–150). Erikson offered a powerful interpretation of the traumatic impact that protracted conflict, forced relocation, and imposed alien ideologies obtained not only on the sociocultural fabric of the groups involved, but also on the private subjectivity and psychodynamics of each group member.

interpretations and conclusions against the backdrop of such phenomenon. I did not try to deny, nor was I ashamed of my feelings. I embraced them as a "natural," human reaction to a difficult and challenging environment, as Devereux posited. I figured that they would help me to better understand how to write properly about what I experienced. I read my overall reactions to my fieldwork experience as the broader countertransferential process that I went through during my stay in Afghanistan. I tried to deconstruct and analyze the feelings of rancor and resentment at being rejected by that very social environment I had so enthusiastically embraced, and the narcissistic injury that I felt burning at the time of my return to the United States. I realized how some negative "judgments" that I had promptly leveled against those among whom I worked stemmed from the "narcissistic rage" of which I was a victim (as Heinz Kohut would put it; see Kohut 1972). My elaborations on such aspects of my emotions helped me achieve a better perspective on the conclusions I had drawn from my experience, and write about them.

Certainly, absolute objectivity is a chimera, yet being aware of one's own biases helps at least to take a step back, and proceed carefully toward a more balanced and accurate account. Indeed, Allen Feldman writes that violence never remains "in a relation of pure externality" to the subjectivity of those who carry out research amidst it (Feldman 1991: 228), and in this vein Parvis Gassem-Fachandi reminds us that the attempt at "extricating" oneself from the pervasiveness of violence within one's subjective states remains crucial "in order to arrive at some degree of objectification" of one's own experience (Ghassem-Fachandi 2009: 4).[15] In reality, I obviously could not blame the people I worked with for behaviors they displayed, which, had I been in their shoes, I would have probably in turn also displayed. This does not mean that I managed to "accept" as appropriate and legitimate, in a moral sense, all that I saw and experienced in Afghanistan, including violence of various sorts—interpersonal and intergroup conflict, domestic violence, and war-related events. Such kind of extreme relativism would amount to falling into the trap that Devereux warned against: the denial of one's own emotions and the power of one's own cultural schemas. We are all inextricably entangled with our own moral and cultural world (perceived sometimes as universally

---

[15] In extreme cases, the scars caused by what one has witnessed and experienced prove to be so deep to render impossible for the ethnographer to adequately report on them, as we can see from Billie Jean Isbell's and Annerose Pandey's accounts of their own fieldwork research (Isbell 2009, Pandey 2009)

valid), and it would be ethically dishonest, and epistemologically fallacious, to deny this.

An equally important dynamic that played out during our interviews was the emergence of unconscious dynamics both in the life trajectory, and the present affects, of the people I was interviewing, as well as the moments of contradiction, redundancy, and backtracking that my informants displayed in session. Associative thinking was a useful resource. In their free-flowing account of their daily lives, and past experiences, my informants associated at times thoughts that apparently had little in common with each other. When I could, I tried to understand the reason for the coupling of two seemingly unrelated topics of conversation, or simple thoughts. The principle is that two streams of consciousness, however unrelated to each other they may seem, are in fact unconsciously articulated verbally together in order to express a third (and possibly a fourth, fifth, etc.), underlying, set of meanings. This was helpful for me in understanding not only what the informant was trying to convey to my attention, but also how this "something" could be put in relation with what we had already discussed together (as well as the informants' relationship with my position as a foreign ethnographer). I elaborate on these instances of associative thinking in the chapters that follow.

Yet, what I felt became even more significant for me and my informants than associative processes was the co-creation of an intersubjective field during our sessions. As mentioned previously, the acknowledgment of the existence of an analytic "field" is possibly the most important advancement in contemporary psychoanalytic theory, shared by more than one school of thought. Though it might seem at first glance that my work with my informants was geared toward making sense of, and understanding, their psychic dynamics over a life span, with a focus on the past, in reality the attention that both put in their emerging narratives was geared toward the here-and-now of each session. Thomas Ogden aptly characterizes this as "the present of the past" (Ogden 2015). I attempted to foster the dialogue between certain self-states in my informants that were possibly still in an unformulated, "non-alphabetized" form (Ferro 2005), and their current subjectivity (Harry Stack Sullivan termed them "not-me" experiences. Sullivan 1953). In other words, I tried to bring to the level of *conflict* those incompatible states of subjectivity that might have been hitherto in a condition of *dissociation* (see Bromberg 2003. I will discuss this process in more detail in the following chapters). My informants were by and large socially functional individuals, who had managed to live through very traumatic experiences and reconfigure their psychological structures accordingly. Yet some of them were still clearly in the grip of personally confusing and destabilizing life contingencies, where their past and current contradictions had not yet fully surfaced to awareness. Indeed, as Donnel Stern notes, "The most crucial events from moment to moment, then, *both inside and outside the consulting room*, are those that resolve the ambiguity of *unformulated experience* into some explicit, conscious shape . . . the factors responsible for resolving that ambiguity are *relational phenomena*, the very events that comprise the analytic field" (Stern D. B. 2013: 640, emphasis mine). I felt often that, though my informants might have managed, with various degrees of success and psychological suffering, to adjust to the stressors and ambiguities in their own lives, the intersubjective field we co-created in our sessions did help them to look beyond the monadic "cocoon" of their immediate self-state and engage those other subjectivities they might have endorsed in the past and be keeping dissociated at the moment of our sessions. From this standpoint, I would say that "psychological growth" happens in that instant of the ethnographic/clinical encounter when the material is told to the ethnographer *not* "as if it were a 'buried' fantasy uncovered by piecing together the links between [an informant's] associations" (Bromberg 2003:707). Rather, Philip Bromberg continues,

> *Therapeutic action is recognized affectively through the process of enactment between* [ethnographer and informant], *where it then has the chance to be symbolized by the verbal meaning attached to the affective perception of what is taking place in the here-and-now. Insight, from this perspective, comes to exist while it is being co-constructed through the interaction of the various and shifting self-states of both* [informant and ethnographer]—*through the interplay between the* "associative content" *and the* 'dissociative context' *that organizes it.* (Bromberg 2003: 707–708)

As such, any "enactment" is an "act of meaning" (Bruner 1990). I think that precisely this capacity to symbolize and alphabetize unformulated experience, acquired through intersubjective exchange in the field created during the session, is what allowed my informants to talk to me and "perceive," make sense of their most important past and present experiences.

1

Historical and Ethnographic Background

The Broader Afghan Picture

The history of modern Afghanistan, since the mid-1700s, is by and large characterized by the dynamics of political supremacy of one of the ethnic groups living in the country, the Pashtuns, over the others.[1] Since Ahmad Shah Durrani, a Pashtun military leader in the service of a Persian ruler, consolidated in 1747 a swath of territory (that only in the nineteenth century would crystallize as modern Afghanistan) into a centralized state, all the monarchical leaders of the country have been of Pashtun descent. When in 1973 the monarchy was finally and definitively toppled and the Afghan Republic constituted, a Pashtun president (Daoud Khan, cousin of the last king, Zahir Shah) took power into his own hands and kept it until he was murdered in April 1978, in the wake of the Saur Revolution that brought the Communist Party of Afghanistan (PDPA) to power with Soviet support. The PDPA members who led the country and fought the popular revolt against their Soviet-backed government from 1978 until 1992 (Nur Mohammed Taraki, Hafizullah Amin, Babrak Karmal, and Mohammed Najibullah) were also all Pashtuns. The Taliban movement that started gaining momentum after the two years of internecine and interethnic fighting (1992–1994), and then eventually took control of the country in 1996 (except for a small sliver of Badakhshan province), was by and large a Pashtun religious revivalist movement strategically organized by the Pakistani military intelligence, whose rank-and-file fighting members mostly grew up in Pashtun refugee camps in Pakistan (Edwards 1998).[2] To this day, the two presidents that Afghanistan

---

[1] Pashtuns represent a plurality in Afghanistan, not a majority. Though there are no reliable census sources to ascertain exact figures (in the past as well as in the present), it is believed that Pashtuns should comprise about 38–42 percent of the total population of the country. Other ethnic groups in Afghanistan are numerous, and none among them approximates the figure usually attributed to Pashtuns in sheer numbers of population.

[2] If it is true that, in the beginning, the Taliban were almost exclusively a Pashtun phenomenon, and increasingly Afghan, nonetheless in the years between 1996 and 2001 individuals from other ethnic groups cooperated as well with the central Taliban authorities, often out of self-preservation and to protect the safety of their own communities. This happened also among Hazaras who, as Shi'a

Crafting Masculine Selves. Andrea Chiovenda, Oxford University Press (2020). © Oxford University Press.
DOI: 10.1093/oso/9780190073558.001.0001

had since the collapse of the Taliban regime in 2001, and under the control of the US-led military coalition, are Pashtun as well (Hamid Karzai and the incumbent Ashraf Ghani).

If there is a single *fil rouge* in the history of the political entity that today we call Afghanistan, I think it can be identified in this ethnic genealogy of power, which has exerted its authority on its subjects mainly in two opposite ways: through strong centralization, and through mild decentralization. The centralizing efforts of the Afghan rulers peaked with King Abdur Rahman Khan (ruled 1880–1901), who worked from within the geographical boundaries that Afghanistan maintains today, and operated ruthlessly to not only consolidate his power internally, but also to subjugate all the pockets of dissent and irredentism that still existed under his control. This meant hitting hard Hazaras (a Persian-speaking, Shi'a Muslim, and ethnically Turkic population) as much as those (Sunni) Pashtuns who did not comply with his demands of total subordination. The central state, in the time of Abdur Rahman Khan, was poised to control closely and micromanage all corners of the Afghan territory. This is a style of governance that Thomas Barfield has aptly termed the "American Cheese model": there will be no "holes" in the reach of the state (Barfield 2010). Yet after Abdur Rahman's charisma and resolve dissolved with his death, his successors realized that it would be too costly and impractical to continue with the American Cheese model and opened up gradually to a "Swiss Cheese model" of governance (Barfield 2010): "holes" in the reach of the state will be tolerated, particularly in those areas that are not economically and strategically crucial for the maintenance of state power. The rulers who followed Abdur Rahman came to terms with the reality of a highly multicultural and fragmented country, which could not possibly be held together by the same one-size-fits-all rules and norms emanating from the center. The Swiss Cheese model proved efficacious until the new centralizing experiment introduced by the Communist Party of Afghanistan (PDPA) in 1978, which led to the final insurrection and then to the disintegration of the state in 1992.

With the advent of the new, post-Taliban Afghan state in 2002, the attempt to impose an American Cheese model of governance has been resuscitated

---

Muslims, were considered by the Taliban as heretics, and as such deserving a capital punishment (Ismail Zakhi, personal communication, 2012. Mr. Zakhi, a Hazara man of about fifty-five years of age, is currently the leader of one of the largest civil society organizations in Bamyan. He spent over one year in a Taliban jail for his criticism of women's conditions under the Taliban government. About the participation of Hazaras in the post-2001, renewed insurgency led by the "Neo-Taliban," see Giustozzi 2008: 48).

once again by the US-backed governments, with outcomes that have been not dissimilar from those obtained under previous governments.[3] In fact, while in the first years after the collapse of the Taliban regime Afghanistan enjoyed a lull in the fighting and conflict that had plagued it uninterruptedly since 1978, in 2004 and 2005 insurrectional activities started developing again against the Kabul government. Not surprisingly, areas of the country inhabited by Pashtun peoples, economically most productive (in the southwest), saw the resurgence of violent opposition against the center. Kandahar, Helmand, and Zabul provinces, with their vast fluvial and fertile plains (which supported also a lucrative illicit poppy cultivation economy, see UNODC 2008), were the first areas to witness antigovernment activity (Giustozzi 2008: 99–106).

Two main factors were at play. First was the resurgence of an anti-foreign, and strongly ethno-nationalist, sentiment among those Pashtuns who had constituted the rank-and-file of the Taliban movement, and who had most benefited from their rule. This sentiment was spurred on by those members of the movement who had not been killed, exiled, or arrested (thanks, in part, to the conspicuous support from pro-Taliban elements in the Pakistani military and political apparatus). Second was the fierce competition, among the "new" military and political powerbrokers, for the exploitation of the resources that the region offered. Many of the "strongmen" that the United States (de facto in charge of the operations in Afghanistan after 2001) chose to bring back to (political) life in order to fight the Taliban in late 2001 had previously been among those *mujaheddin* who fought the Soviets in the 1980s (also supported by the United States, among others). After the Soviet withdrawal, and the collapse of the Afghan Communist government in 1992, these men became the protagonists of the gruesome civil war that followed, which laid waste of the whole country and heralded (if inadvertently) the birth of a revolutionary, revivalist movement: the Taliban (Giustozzi 2009: 69–85). The Taliban, starting from Kandahar province and proceeding swiftly to the rest of the country, were at first welcomed with relief for their capacity to put an end to the chaos and lawlessness that had enveloped the

---

[3] Indeed, Thomas Barfield contends that the inability of the US political and military apparatus to apprehend this simple lesson based on the historical precedents in the country, coupled with their obsession with the "golden rule" of allowing only Pashtun leaders to sit at the helm of the country (instead of facilitating a more culturally sensitive passage into a federalist system, which is after all also the American model of governance), has paved the way for the destabilization of the country that we are still witnessing in 2019 (Barfield 2010).

country after 1992, at the hands of the myriad big and small strongmen who brutalized the population on a daily basis, mostly for their own benefit.

In Pashtun areas of the country, the former *mujaheddin* leaders who had been ousted by the Taliban were given a new chance in 2001 and 2002 by the US military and strategic decision-making apparatus (Giustozzi 2009: 87–90). These men, who had been politically and militarily (and sometimes physically) obliterated by the Taliban, were put in charge by Washington of the political and security control of the areas that the Taliban were gradually losing to the international troops and their Afghan supporters. They reassured their international (mainly American) advisers that they still retained their own popular base and constituency, and that they could provide stability to the areas left without viable authority figures by the demise of the Taliban regime. While in the first years after the fall of the Taliban government it seemed that some stability had been regained, in reality the former *mujaheddin*, now also institutionalized political leaders, worked to build fiefdoms for themselves from which they could extract resources and organize illicit economies (Nader Naderi, personal communication, 2009).[4] While at times earnestly engaged in military struggles with a resurgent Taliban movement (the "Neo-Taliban," as Antonio Giustozzi termed them, see Giustozzi 2008), the strongmen also took advantage of their institutional position of authority, and their connections to the international military apparatus, to play a double-sided game with the insurgents, exploiting the latter's close ties with specific rural communities. Illicit economies based on opium poppy cultivation, as well as heroin production and exportation, thrived in the southwestern provinces more than they did in the rest of the country, where these activities had been an important source of livelihood for a long time (UNODC 2008). Not surprisingly, the rise in economic and political power that the strongmen enjoyed went hand in hand with an intensification of the insurgents' military operations in the same provinces. At the height of this situation, in late 2009, the US government decided to increase the number of troops operating in the southwest of Afghanistan, and implement a stronger military policy in the region (the so-called surge). It has widely been recognized as an accomplishment of President Obama's administration that the intensity of the insurgency in the southwest was curtailed

---

[4] Nader Naderi was until 2011 the head of the Afghanistan Independent Human Rights Commission. For a detailed study of the role and history of the so-called warlords in Afghanistan, see the excellent articles by political scientist Antonio Giustozzi (2003, 2006), and his 2009 volume, *Empires of Mud: War and Warlords in Afghanistan*.

by the "surge." And indeed, the operations of the anti-government and anti-NATO forces did measurably decrease in the region (Chivers and Filkins 2010). What is less acknowledged, however, is that, soon after this shift of events, insurgent activities increased elsewhere, namely, in the southeast. What most likely happened (though there is no definitive evidence for it), is that, confronted with an overwhelming NATO military effort, the insurrectional forces quietly moved their fulcrum of operations to an area that had remained relatively peaceful, and therefore somewhat unguarded. The provinces of Nangarhar (where I conducted my fieldwork), Kunar, Laghman, and Nuristan bore the brunt of this strategic shift on the part of the insurgents. By 2009, when I reached Jalalabad for the first time (Jalalabad is the capital of Nangarhar province), the situation had already started to deteriorate, and would continue to do so at an increasing pace until I left the field in 2013.

By moving from the southwest to the south-east, however, the insurgency also morphed into something slightly different from what it had previously been. The socioeconomic environment wherein Pashtun populations live in the south-east is radically different from that in the southwest. Scarcity of natural resources and rugged terrain meant that the Pashtuns of the southeast (collectively known as Ghilzais, as opposed to the Durranis of the southwest) never developed the stratified, hierarchical social structure that could be found in the fertile plains of Kandahar, Helmand, and Zabul provinces (Barfield 2014). Pronounced segmentation/fragmentation into nested social groups (traditionally categorized in Western scholarship as tribes, lineages, and families—in Pashto *qawm, khel,* and *koranay,* respectively), and a cultural ethos of strong egalitarianism and individualism, made a marked impact on the organization of the insurgency as it moved from its stronghold in the southwest to the mountains of the south-east. Even assuming that the insurrectional movement had ever existed as a truly centralized entity in the southwest (which is yet to be fully demonstrated), certainly it took a fragmented and "anarchoid" character in the south-east. During my fieldwork in Nangarhar province, speaking of "Taliban" had already become a general, almost conventional shortcut to signify simply "rebel" or "insurgent." None of my informants, including those who lived in very volatile and unstable districts of the province, knew exactly what the term "Taliban" meant any longer. They were unable to discern any unity of command, composition, or aims in the military operations that were carried out against the Afghan and the NATO troops. It was also common, I was told, that the inhabitants of an area struck by a militant attack, whether against military or civilians, would not know where to turn in order to find and negotiate with the responsible group, in the effort of amending the possible frictions that could have led to the attack, and preventing future ones. Multiple groups were at work, my informants explained to me, each with different and sometimes incompatible motives, made up of both local and foreign elements, and possibly supported logistically by outside powers (such as Pakistan, for example, see Giustozzi 2008: 21–23).[5] Some groups, in my informants' view, seemed to have retained that religious and puritanical spark that had characterized the Taliban movement in the 1990s. Others emphasized in their contacts with the populace more ethnonationalistic overtones, stressing the need to preserve a strong Pashtun cultural component at the helm of the Afghan state. Still others appeared to my informants as composed by little more than violent bandits, moved largely by hopes of personal gain and political rewards. Some groups claimed affiliation to a "Taliban" movement, others claimed to be closer to other insurgent factions (like the long-lived Islamic party Hizb-i-Islami Gulbuddin, or the more recent development of the Haqqani network from Khost province), while others did not even bother to declare any form of broader affiliation.[6] All, however, shared similar guerrilla warfare tactics in combat, and targeted the same representatives of the Afghan institutions (both civilian and military), and the international military forces. The uncertainty of the situation created a sense of constant anxiety, fear, and helplessness in most of those with whom I had relationships.

---

[5] The most common claim among my informants was that Pakistani Pashtuns were in large part to blame for much of the disturbance occurring in the area. ISI, the main Pakistani intelligence agency, was also considered to be the puppeteer for some of these "intruders." Some mentioned the presence of Arab, Caucasian, and Central Asian fighters hiding in the mountains (supposedly it was possible to recognize a Central Asian militant by the fact that his dead body did not show signs of circumcision). How much of all this was fantasy, and how much reality, is hard to tell. Certainly, the presence of Pakistani Pashtun militants in the southeast has been often confirmed by Western intelligence sources, and so has the supporting role of ISI. The same, though to a lesser degree, can be said of Arabs, Caucasians, and Central Asians, who are believed to train in the camps set up in the FATA, only to pour into Afghanistan at a later time. I myself have often seen groups of young Mehsud and Wazir Pashtuns from FATA wandering the bazaar's alleys in Jalalabad (their clothes and haircuts betrayed their origin). How much weight and relevance these foreign elements held in the broader economy of the insurgency is very difficult to gauge.

[6] Gulbuddin Hekmatyar and Jalaluddin Haqqani are among the few "survivors" of the major original *mujaheddin* commanders who received enormous amounts of money and equipment from Western and Middle Eastern countries with the expectation that they would organize the resistance against the Soviet army and the Afghan government. Since the beginning, they displayed an extreme religious fervor and fighting spirit that was highly appreciated and encouraged by their Western backers. Thirty-eight years later, both in their seventies, Hekmatyar and Haqqani have turned that same fervor against those who once had supported them—the Western powers.

The fieldsite in perspective

Jalalabad, the capital of Nangarhar province, is historically one of the intellectual centers of Afghanistan. Aside from hosting the second-oldest university in the country (after Kabul University), it has been alongside Peshawar, Pakistan, the center of northern Pashtun culture (the southern fulcrum was Kandahar). The Pashtun groups (traditionally termed "tribes") who to this day inhabit the area are common to both the Pakistani and Afghan side of the international border (a border that was imposed in 1893 with the name of Durand Line, through an agreement between the then-king of Afghanistan, Amir Abdur Rahman, and the British Raj). In fact, I discovered during my fieldwork that the cultural, social, and economic bonds that tie Peshawar to Jalalabad are much stronger than those that tie Jalalabad to Kabul. A town of sizable proportions (around 250,000 inhabitants at the time of my fieldwork), Jalalabad is the only such center between Kabul and Kandahar. For decades, the city has also been the locus of governmental visibility and administration. It is not a coincidence that Jalalabad was in fact never militarily "conquered" by the anti-government and anti-Soviet *mujaheddin* in the 1980s and early 1990s, but rather withstood all their assaults, and capitulated only when the local military garrison of regular Afghan troops decided to relinquish control over the city after the collapse of the last Communist government in 1992 (Nojumi 2002: 95–98).

As it happened in most areas of Afghanistan, the 1992–1994 civil war hit Jalalabad in a devastating manner. It is not surprising then, that when the Taliban made their appearance in Nangarhar province, in 1995, they took control of the capital without much fighting, and were said to have been welcomed by the local population as the only ones who could effectively rid the province of the plundering *mujaheddin* militias.

During the Taliban regime, Nangarhar province, as well as those surrounding it, enjoyed a period of peace, albeit amid economic struggle. Opium poppy cultivation, although officially forbidden by the government, was employed regularly as a cash crop by farmers on a small scale (differently from what happened in the southwestern provinces, where it amounted to a large industry geared toward exportation. See UNODC 2002). Families owning large swaths of land were rare, as had always been the case, and the latifundium model of production that was common in Kandahar and Helmand provinces (with its related relationships of quasi-serfdom for the landless laborers) was rare in Nangarhar (Kakar 1979: 188). During the fall of the

Taliban regime, Jalalabad and most of the region were spared the worst of the fighting. The aerial assault mounted by the international coalition was concentrated instead in the mountainous areas in the southern part of the province, on the border with the Kurram and Khyber Agencies of the Tribal Areas of Pakistan (FATA). Unlike Kandahar city, which fell to the international coalition after fierce clashes, Jalalabad experienced a smooth transition to the provisional government sponsored by the international community.

Nangarhar province is part of a broader rural region. Cut deeply by the Kabul river in a west-east direction, and by the Kunar river (a tributary of the Kabul river) running north-south, Nangarhar and the surrounding provinces enjoy verdant and fertile plains along the main streams of water, while presenting a dryer landscape increasingly devoid of any vegetation the further one ventures away from the rivers. Small-scale agriculture, which is the main economic activity in the fluvial plains, is almost completely unfeasible in the arid and rocky soil of the rest of the region (Khaurin 2003: 4–12). Agriculture (where possible) is carried out in fields that can be either artificially irrigated (abi) or irrigated by rain (lalmi). Artificial irrigation is obtained in three ways: by channeling the water streaming down from a mountain slope, by channeling the water from the Kabul and Kunar rivers, or by a system of canals (karez) that bring water from the mountains down to the fields in the adjacent plains via underground tunnels (Kakar 1979: 184–187). The latter system has not changed much during the decades, as it was witnessed and described in detail by Anderson (1978a), and confirmed by my personal observations. Karezuna (pl. of karez) have existed for hundreds of years in Afghanistan (Kakar 1979: 184–187). Modern karezuna were extensively put in place by the Soviet-backed government in the late 1970s and 1980s, together with works of restoration carried out on the already existing tunnels. It is mostly through the effectiveness of these recent karezuna that intensive cultivation is possible to this day in the dry plains surrounding the two main rivers.[7] Main crops include wheat, maize, barley, cotton, alfalfa, and poppy, as

---

[7] Interestingly, several of my informants complained to me, almost in disbelief, about the lack of infrastructural commitment that the international forces demonstrated during the twelve years (as of my fieldwork) of their occupation of Afghanistan. These informants could not understand how their "enemies" (the Soviets) managed to actually show some sort of concern (as perceived by them) for the welfare of the rural population in the area, by building large infrastructural ameliorations (such as canals, power plants, roads, and residential compounds), while their "friends" (the US-led coalition) showed a complete lack of interest for the same issues. Those among my informants who were not sympathetic to the international intervention in the country (or those who had over time become so) interpreted the contrast as the apparent proof of the lack of good faith on the part of those who were leading the international campaign in Afghanistan.

well as various fruits and vegetables. However, the flat and cultivable strips of land adjoining the two major rivers soon become rugged and semi-arid mountainous terrain, rising up to the 15,620 feet of Mount Sikaram, within the Spin Ghar range in Nangarhar province. During the decades of war the luxuriant coniferous forests that once covered the mountainous areas in the provinces of Nangarhar, Kunar, Laghman, and Nuristan have been incrementally cut down by the local population to supply firewood and construction materials (Delattre and Rahmani 2007). Wildlife, once comprising several species of ungulates and their predators, has been harvested for food, dramatically decreasing its numbers (Karlstetter 2008, Stevens et al. 2011). The population in Nangarhar province, who are largely Pashtun and Sunni Muslims, lives mainly clustered in small villages, whose inhabitants usually share close kinship ties. Large areas of the province, particularly in the south, are sparsely populated. The only "legitimately" urban area is the capital, Jalalabad. The rural districts' centers are often little more than trading hubs for the population who live in the districts' villages, and the centers for the thin administrative structures that the state offers in each district. People from the districts tend to congregate in the district's center in order to exchange or buy goods, address the local authorities, and meet acquaintances and relatives from other, more distant villages. Inhabitants of the dryer areas carry out mostly a subsistence type of agriculture (where possible at all) and maintain small numbers of domesticated animals, also on a subsistence basis, and for barter.

Pashtuns in ethnographic perspective

Pashtun peoples, both in Afghanistan and Pakistan, have received the attention of scores of ethnographers and anthropologists in the past. These include, among others, Fredrik Barth (1959, 1966, 1969), Charles Lindholm (1982, 1996), Akbar Ahmed (1976, 1980, 1991), Willi Steul (1980), Bernt Glatzer (1977), Jon Anderson (1975, 1978a, 1978b, 1983, 1984), Benedicte Grima (1992), Inger Boesen (1980, Boesen and Christensen 1982), Nancy Tapper (1973, 1977, 1991), and more recent analyses by Bernt Glatzer (2002) and Thomas Barfield (2008, 2010, 2014).[8] In these works, the

---

[8] Among the very few Pashtun academics who researched and wrote in English about Pashtuns are the Afghan anthropologists Alef Shah Zadran (1977) and Ashraf Ghani (1978), and the Pakistani legal scholar Haroon Baryalay (2005, 2006). Only the work by Ghani has been published. Ashraf Ghani is now (2019) the president of Afghanistan.

ecological settings, social structure, and various cultural productions of Pashtun human groups inhabiting both sides of the Durand Line have been expounded in detail.[9]

These accounts sprang from the observation of a shared sociocultural environment (the Pashtun milieu), immersed into diverse political, economic, and ecological realities. For instance, the Pakistani Yusufzai Pashtuns whom Barth and Lindholm studied in Swat lived in a context radically different from that in which the Afghan Ghilzai Pashtuns studied by Anderson in Khost lived (albeit in the same historical period, in the case of Lindholm's subjects). Such differences in context might have been strongly pronounced between Afghan Ghilzais and Pakistani Yusufzais, but less so between Afghan Ghilzais and the Pakistani Mohmands whom Akbar Ahmed researched. These context-dependent peculiarities, however, have apparently developed from a common "bedrock" of cultural material (idioms) and basic social arrangements that were common to the Pashtuns analyzed by past ethnographers. Upon the basic nature of this "bedrock," in fact, there was general agreement. I too found my fieldsite to present characteristics, in terms of cultural idioms, not dissimilar from those reported by previous scholars. Pashtun society at large appeared to me, as it did to earlier investigators, to be a strongly honor-based society, where the concept of public shame is equally strongly present. Patrilineally and patrilocally structured, it presents

---

[9] Of course, this list is by no means exhaustive. I mention here only the major ethnographers and anthropologists who have worked on Pashtuns, and whose work I feel proved most influential for subsequent research. Additionally, it must be remembered that there exists a ponderous colonial literature (mostly British) about the areas now encompassed by Pakistan and Afghanistan, and in particular about Pashtun people. Since the explorative journeys of Mountstuart Elphinstone into Afghanistan on behalf of the British Crown, at the beginning of the nineteenth century, until Olaf Caroe (the very last governor of the North-West Frontier Province in 1947), there have been countless accounts (official or private) on the people who live in the area straddling Pakistan and Afghanistan. Charles Lindholm, in a 1980 essay (Lindholm 1980), makes a case for the usefulness, to the contemporary scholar, of the colonial "ethnography." He points correctly to the sharp differences in the kinds of personal mindset and motives that lie behind each account. Whereas Mountstuart Elphinstone's and Charles Masson's, Lindholm argues, represented the reports of individuals relatively removed from the political machinations that followed, others, like Alexander Burnes's (the first British victim of Pashtun irredentism, in 1841 or 1842), and Olaf Caroe's, incarnated a relationship between parties that had already entered the phase of power struggle, and developed into full political control after the establishment of the British Raj. The diatribe, within anthropology, about the meaning of the "colonial encounter" has filled volumes (for the most conspicuous of which, see Asad 1973), and it is not my intention to participate in it. It is interesting to note, however, that to this day the bookshops of the main bazaar in Peshawar (Sadar bazaar), and the only extensive bookshop existing in Kabul, are still replete with the modern reprints of the many colonial texts that were published over time, from the most well-known names (as the ones I mentioned), to the most obscure military gazetteers and bulletins, published for internal use of the administration during the colonial rule (of which I myself gathered several specimens). What this tells us about the state and ramifications of the "colonial encounter" (which is, in some regard, still ongoing) would be worth investigating.

a heavily enforced segregation between sexes, whereby any kind of relationship between two people of opposite sex who are not closely related by blood or marriage is avoided as shameful. As it is the case with many ethnographic examples throughout the Mediterranean and Middle Eastern regions (see, among many examples, Bourdieu 1977, Giovannini 1981, Herzfeld 1985), among Afghan Pashtuns as well the behavior of women is taken as the yardstick for a judgment on the respectability and honor of a family as a whole. "The women of my family are my *namus*," you often hear men say, which means that their honor, and that of their family members (including the women's), rests on the appropriate behavior of their women.

It is worthwhile reiterating here the importance of the patrilineal structure of Pashtun society. Descendance or ascendance within the lineage is traced down or up through the male line only, and indeed true lineage mates, in the full sense of the term, are only blood relatives on the *male* side (including women—for example, one's father's sister, but not one's mother's sister). This state of affairs has significant repercussions in gender ideology and interpersonal behavior. Within a Turkish context, Michael Meeker has defined *namus* as "sexual honor" (Meeker 1976: 260), by which he means that "a man's 'reputation,' 'social standing,' or 'social legitimacy' is tied to the chastity of specific women" (260). For Meeker, namus refers to a "state," which can belong to the collectivity or to an individual person, and which one either has or has not, without degrees or nuances. In the same vein, Diane King adds that "Namus is a symbolic attribute of a patrilineage" (King 2008: 319). Insofar as namus depends on the behavior of women, and patrilinearility confers to men the exclusive privilege to be the sole active perpetuators of descendance, it is incumbent on the men of the family to ensure that their family's women will behave appropriately to maintain the family's namus intact, above public reproach or even suspicion. In their Turkish and Iraqi Kurdish contexts, Carole Delaney and Diane King, respectively, have tied this principle to what they term "monogenesis" and "patrogenesis." By this they mean that in both their fieldsites, the belief shared by women and men alike was that only men participated actively in the passage of biological traits of any sort to the next generations. In other words, "the child originates with the father, from his seed" (Delaney 1986: 497), while women provide an undifferentiated "soil" in which the seed will grow, and which will not affect any of the characteristics of the final product, the child. From this standpoint, Delaney has likened the woman's body to a field, which can be fertile or barren (Delaney 1986: 498). The field must be fenced off, protected from any intrusion and interlopers, who might plant their seed unbeknownst to the owner of the field itself (i.e., the patriliny of the woman's original family, first, and of her husband's, later, for obvious reasons). Diane King, building on Delaney's theorization of the woman's body, wrote of namus as strictly related to *sovereignty*—the sovereignty by the woman's patrilineage over her body (King 2008: 319). King finds a causal link between the sovereignty of the woman's patrilineage over her body and the protective and controlling behavior over the woman's mobility and public life enacted by her family (including by the family's women, who also police the younger ones). Anger and rage erupt among the women's male blood relatives when they are given the opportunity to even vaguely suspect, not to mention demonstrate, the sexually inappropriate behavior of one of their family's women. The result is often violence against the woman (and occasionally also against the man involved—a so-called honor killing (King 2008: 324–325. See also Ring 2006).

Despite the strong patrilineal and androcentric structure of Pashtun society, a patrogenic or monogenic belief, in King's and Delaney's terms, was not maintained by many of my informants (particularly the younger ones). I believe this was because of mainly two reasons. First, in the past fifteen years, Afghans have been exposed to intense interactions with the international community present in their country, both military and civilian, which has aggressively advanced an ideological agenda based on Western human and civil rights. Most Afghans, whether in agreement with it or not, are aware and discuss such narrative. In addition to this, we have to account for the deep reach of global information networks such as the Internet, which many Afghans now can access from their phones, particularly in urban areas but increasingly also in rural ones (the widespread lack of electricity does not allow the reliable usage of laptop or desktop computers, but smartphones connected to the Internet are not uncommon, often charged by car batteries in dedicated shops in bazaars and markets).[10] Second, since 2001, and partly also before then, millions of refugees have slowly found their way back to their original Afghan districts and villages from the neighboring countries, where they spent many years after fleeing the conflicts. As the Canadian-Iranian

[10] Let us not forget that a dependable network of cell phone communications in Afghanistan after 2001 was one of the first things that quickly materialized in the wake of the US-led occupation of the country. This move had its main reasons in the need of the military and civilian international personnel to communicate with each other easily and reliably, as well as with their Afghan counterparts. The Afghan population took ready advantage of the new technology. The same I saw happening in Iraq in 2003 and 2004 (where I was working as a freelance journalist) after the US-led invasion of the country toppled Saddam Hussein.

anthropologist Homa Hoodfar recently remarked, for many returnees "New economic conditions, and the refugees' experiences of life in other Muslim contexts, have fostered new understandings and alternative visions of gender roles and of 'Muslimness.' Views on women's education, women's participation in the labor market, and women's rights in the family have shifted substantially, in part because of dire financial necessity" (Hoodfar 2009: 224). To these two main reasons for the potential disappearance of a monogenic/patrogenic ideology, one must also add the pervasive Soviet influence in the country, which, from 1973 until 1992, held sway in the educational curricula of most public schools and universities in Afghanistan, with its heavily rationalistic and anti-traditionalist slant. Also, tens of thousands of rural and urban Afghans were sent to the Soviet Union on scholarships to attend Soviet schools and universities.

In the end, it was not possible for me to ascertain whether, prior to the unfolding of the multiple Afghan conflicts since 1979, an ideology of patrogeny/monogeny might have been present among the population.

Apart from the uncertainty about the presence of a patrogenic/monogenic ideology at the basis of gender relations in the context where I worked, the other characteristics of Meeker's, Delaney's, and King's model do by and large apply to my Pashtun case as well. Descent and ascent are strictly traced through the male line only. Indeed, if the family's women must, through their "public" performance, ensure the maintenance of the family's honor and public respectability, it is upon the men to protect, and restore, such respectability when it becomes tainted by some delict or impropriety on the part of the women, or by a man outside the family toward its women.[11] Usually such a taint entails some violent retaliation, either against the family's woman, or the outsider offender, or both, if they are deemed to have acted in accord. A male family member who has *ghairat* (who is *ghairati*, or, alternatively, who is a *ghairatman*; see below for details) must take action in these cases, lest he be publicly disgraced and considered *beghairata*, that is, unworthy of

---

[11] Paradoxically, the "public" performance of Pashtun women outside Kabul (where practices are more "relaxed") must be intended as a "negative" performance, in that the less they access the public sphere, the more virtuous and respectable they will be considered, and by extension the namus of the family as well. Unlike the women in Diane King's ethnographic context, who were allowed a degree of public interaction with male non-relatives, and even outside mobility (albeit closely policed), Pashtun women in rural contexts will stay mostly at home and interact only with blood relatives also when at home. Pragmatic exceptions to this rule, however (e.g., for women working in aid and development agencies), are becoming allowed recently in urban areas like Jalalabad, in cases of severe economic hardships suffered by the women's families in the wake of the devastation caused by the continuous conflict. See Chiovenda M. (2012).

being a Pashtun, of being a *nar* (a "manly," virile man; see).[12] The equivalent of namus in the male realm is *izzat* (what Meeker would call "sharaf." Meeker 1976: 243–260). A man's izzat is injured when the man sees his rights disrespected (for instance, by being cheated in an economic transaction), or is publicly insulted, or is not accorded the esteem, as a nar, that every Pashtun man, regardless of its social status, wishes to claim. As in the case of namus, so also in the case of injury to izzat, the way to restore its integrity often entails violent retaliation, which every man who calls himself ghairati is required to enact. Failure to do so carries with it the stigma of being beghairata, and the loss of personal respect within the community at large. Ethnographic literature concurs with such general portrait of Pashtun society, in which these cultural idioms and social arrangements represent a central value.

One concept that is pivotal for the development of this book, and that is not discussed in depth by the previous literature, is *nartob*. Nartob, in a nutshell, refers to the moral and ethical qualities that a nar must possess. A nar is not just an "ordinary" man (*saray*, in Pashto), but rather a "manly man" (the suffix -*tob* in Pashto has the same meaning as the suffix -*ness* in English. Thus, nartob is the exact equivalent of English "manliness," or even more closely from a semantic standpoint, "virility"). Indeed, Michael Herzfeld's felicitous definition for his Greek Cretan context is here even more apt: a nar has "to be good at being a man" (Herzfeld 1985: 16). Strictly related to nartob is *ghairat*, which is the willingness and capacity to demonstrate publicly that one is a nar. So, a nar is a virile man, a man in all his masculine prerogatives, which correspond with certain characteristics that are often associated with virility in many other ethnographic contexts—courage, strength, fearlessness, assertiveness. Yet nartob has also another side to it, which is not directly linked to

---

[12] This is obviously true and inescapable in the case of the original family of an unmarried woman: her behavior reflects on the reputation of her family (i.e., her father, brothers, and then mother and sisters). After a woman marries (or, more precisely, is married off to someone, for which there exists in Pashto language a passive verb construction, as opposed to the active verb construction that describes the act of marrying for a man), the responsibility over her behavior, and the consequences of it, become somewhat less clear-cut. Opinions among my informants varied as to who had the major responsibility to act in order to restore the honor of one's family, should the married woman engage in any reproachable behavior (mostly related to proper inter-gender relationships, not necessarily involving sexual intercourse, whether suspected or flagrant). The original patrilineage of the woman invariably would suffer a blow because of her behavior. Her husband, however, would also in turn be considered *beghairata* (unmanly, see below) if he did not take action against her. This would definitely reverberate on the honorability of the husband's family as a whole, prompting his brothers and father to pressure him into some sort of action. I was told of cases in which the initiative was first taken by the brothers of the married woman, or alternatively by her husband, or directly by his own brothers. For more on the patterns of violence among Pashtuns as portrayed by previous ethnographers, see Lindholm 1981a, Ahmed 1980: 126–160 and 181–213, Barth 1959.

aggressiveness and violence. Protecting the rights of the weak, supporting the poor, defending women and children, awareness and cognizance of one's *qawmi laar* (the ways/customs of one's tribal group of ascription), diplomatic skillfulness in managing instances of social friction and conflict, are all expected in a nar. Obviously a nar is also someone who possesses and displays ghairat, which is the capacity to act promptly and effectively in cases in which one's namus or izzat have been compromised by another person's lack of respect or improper public behavior. Most of the time, when the public display of ghairat comes into play, it is in situations requiring retaliatory action, which are usually violent to some degree. Still, violence and aggressiveness per se are not necessarily the primary attributes of a nar. There exists, so to speak, a "good" violence, and a "bad" violence. One of my informants, a twenty-six-year-old man from a rural village near Gardez, in Paktia province, told me:

> You know, there are two kinds of ghairat. There is good ghairat, and bad ghairat. For example, imagine you are walking in the street with your sister, and a strange man passes by and looks at your sister intensely, as if she knew her. If you start speculating that they are having an illicit relationship, and kill them both without even inquiring with any of them, you are not a ghairati man, you are ignorant and stupid [besauada aw kamaqal]. On the contrary, if you discover that your wife has been the object of bad verbal or physical harassment against her will, and you kill the culprit, this is good ghairat, something obligatory [majbur].

In other words, violent behavior that is carried out as the proper expression of a man's ghairat in defense of his namus or izzat is not only "good," but strictly necessary, lest he be branded a beghairata (lacking manly attributes, in colloquial terms). Simultaneously, angry and unjustified acts of violence, which go against the principles upon which the idea of a nar rests, also render the perpetrator beghairata. Abuse, bullying, and other behaviors that purposely offend the namus or izzat of others are themselves shameful.

Notwithstanding the caveats that emerged from the material gathered during my own research, it is undeniable that the idea of revenge (*badal*), as the means to wash away a stain to one's own, or one's family's, honor, is a predominant cultural element in any Pashtun social environment; its presence remains pivotal for the social status of the individual, whatever his personal attitude toward revenge may be. Other shared values of Pashtun culture that previous ethnographers underscored (for example, the formal recognition accorded to the practice of hospitality [*melmastia*], the right of any man to seek refuge in the house of any other man [*nanawatay*], and the institution of an ad hoc council of elders [*Jirga*] in order to adjudicate cases of interpersonal or interfamilial conflict) remain generally relevant to this day, as I was able to corroborate during my fieldwork.[13]

Yet it is of the utmost importance to remember that these aspects represent only *cultural idioms*, which tell us about the "raw" material with which the individual's subjectivity has to work, as a social and cultural being, They *cannot tell us much*, though, about how the individual himself inhabits these idioms, and how they have an impact on his psychological dynamics.[14]

Additionally, we should always keep in mind that almost four decades of traumatic and disruptive events have taken place in Afghanistan to this day: a popular insurrection, though heavily Western-instigated and -backed, against the Soviet invasion and the local government it supported (1978–1992); a devastating internecine and brutal civil war (1992–1996); a short-lived and obscurantist, if religiously revivalist, political state system (the Taliban regime, 1996–2001); and, finally, the new military US-led occupation that prompted a resurgent popular uprising (2002–present). Forty-one years of practically constant war and destruction enveloped Afghanistan, and indirectly Pakistan as well, since Charles Lindholm wrote his ethnography in 1982. Enormous (and traumatic) social changes and cultural shifts would be legitimately expected to have taken place due to this tumultuous history (as in fact I found was the case). This simple fact alone forces the contemporary ethnographer to consider previous literature only as a general compass, unless the aim of the research is specifically historically comparative.

We will see how real individuals in times of severe disruption caused by violent conflict subjectively deal with, reject, transform, and maneuver around

[13] Let us also not forget that many of the cultural schemata and social arrangements related to masculinity, and manliness, that Pashtuns by and large publicly display, are certainly not unique to Pashtuns alone. There exists an abundant ethnographic literature about "honor and shame," patriarchal structures, and female sexual purity, which indicates that similar sociocultural patterns and moral beliefs about how "to be good at being a man" (Herzfeld 1985: 16) can be found in very different areas of the world, from South Asia to North America (see, among many others, Peristiany 1966, Gilmore 1987, Giovannini 1981, Herzfeld 1985, Nisbett and Cohen 1996, Blok 1981, Aase 2002, Johnson and Lipsett-Rivera 1998).

[14] Concerning this point, Melford Spiro wrote that "we have for too long—certainly since Durkheim—accepted the coercive power of cultural symbols on the human mind to be a self-evident truth" (Spiro 1982: 45).

shared but ever shifting social arrangements, cultural values, and idioms, and in so doing create new models, which may usher in (if imperceptibly) social and cultural change. Their effort was rendered more complex and, sometimes, painful precisely by the shifting conditions that their sociocultural milieu was undergoing because of the conflict.

Again, as Judith Butler noted long ago, every time a behavioral or conceptual paradigm is (apparently) repeated, it also is necessarily disarticulated and recomposed in a slightly different shape (Butler 1988). There is never full duplication in repetition.

"Culture Talk," war, and subjectivity

In the wake of these latter considerations, let me point out that two contemporary and influential anthropologists, Mahmood Mamdani (2005) and Lila Abu-Lughod (2013), have recently raised the issue of how "culture" is (mis) used in scholarly and policy-oriented explanatory models, and turned into an easy scapegoat for much of what the "West" does not, or does not want to, understand of social and political dynamics among peoples today strongly impacted by the strategic and military policies of Western powers. And digging a little deeper, I dare to say that it might not even actually be a matter of "not understanding" after all, but instead a convenient, and to a degree conscious, way of misrepresenting the "Other," so to conceal one's own material and moral responsibilities in the ongoing catastrophic circumstances that apply currently to vast areas of the greater Middle East. In fact, referring specifically to Afghanistan, Abu-Lughod asks: "Why was knowing about the culture of the region—and particularly its religious beliefs and treatment of women—more urgent than exploring the history of the development of repressive regimes in the region and the United States' role in this history?" (Abu-Lughod 2013: 31). Mamdani calls this approach to the Other, and in this particular case the Muslim Other, "Culture Talk." He writes: "Contemporary Culture Talk dates from the end of the Cold War. . . . It claims to interpret politics from culture, in the present and throughout history" (Mamdani 2005: 20). In pointing out the pitfalls entailed by this stance, he adds: "Culture Talk does not spring from the tradition of history writing but rather from that of the policy sciences that regularly service political establishments" (27). Afghanistan is the focus of a vast section of Mamdani's book as well.

I acknowledge the validity of Mamdani's and Abu-Lughod's critique, and do not use "culture" in this book as an explanatory paradigm for all the behaviors and inner vicissitudes my informants suffered during their lives— certainly not to "explain away" the severe structural, economic, and social problems that Afghanistan, as a whole, has accumulated over the course of the past four decades. It must be clear from the start that, not surprisingly, forty-one years of continuous conflict did have a devastating, catastrophic outcome on how Afghans live their lives daily. This reality cannot be overstated enough. This book builds upon the implicit understanding not only that cultural material and idioms of masculinity are in constant flux, and never static, but that in Afghanistan they have been dramatically impacted by the routinization of disproportionate violence that obtained in the wake of the decades-long state of conflict. Elsewhere (Chiovenda A. 2018a) I have detailed the specifics of what I think was a process of routinization and institutionalization, ongoing to this day, by which the moral grounds for performing "just" violence according to Pashtun cultural principles have been slowly and significantly eroded, and modified. As a result, the principles themselves that now regulate culturally appropriate intracommunity violence have a very different outlook, according to my informants, than they held, say, in the late 1970s. I touch upon this particular shift in the perception of what is legitimate violence in multiple passages of the book, because it emerges strongly from my informants' narrative itself.

All my informants' sense of masculine worth, and their struggle to affirm a personal and culturally acceptable masculinity, owe something, in various ways and to different degrees, to the fact that a violent conflict has engulfed their country since the day they were born. This figures prominently in their narrative and mindset. As we will see, for example, Rohullah is adamant in recognizing that the rural village where he moved to escape the civil war in Kabul, when he was about six years old, owed much of its intolerable proclivity for interpersonal conflict to the misery that the Soviet occupation and destruction, as well as the civil war, brought upon it. Likewise, Umar, a committed Talib preacher until the start of the US-led aerial campaign over Nangarhar province, knows that his early personal profile owes to the indoctrination into an anti-Western politico-religious rhetoric that had its roots in the rejection of imperial policies of the US and major Western countries. Furthermore, his life-changing "rehabilitation" happened thanks to his immersion into the new and unknown world of Western aid organizations that poured into Afghanistan right after the fall of the Taliban regime.

A new avenue of expressing his own masculinity emerged and was endorsed by him. Finally, Baryalay lives on a daily basis in a rural village that is completely overrun by the fight between the insurgency and the Afghan military and police. He lives shoulder to shoulder with individuals whose ideological outlook, and behavioral expressions, are strongly skewed by the violent conflict they have literally at their doorstep, and which in turn is the direct result of the occupation and country-wide military campaign that the US-led coalition has carried out in Afghanistan for more than twelve years (at the time of my fieldwork research). He knows this and he says so.

These brief mentions will suffice to anticipate an aspect of the book the reader will find articulated in more detail in the following chapters. Mahmood Mamdani reflects back at the early days after September 11, 2001, when he "had the impression of a great power struck by amnesia" (Mamdani 2005: 15). Despite my focus on the interaction between cultural, material, and social arrangements on the one hand, and psychic processes and phenomena on the other, my aim in this book is not to perpetuate that amnesia and depoliticize human misery and violence. I am all too aware of the pernicious consequences that the strategic and military policies of the United States in Afghanistan, and elsewhere in the Muslim world, have had on millions of people in the past four decades, just as I am reminded of the equally pernicious contribution that other countries, bent on their own strategic interests, imposed on the same populations (to mention but the main ones, the Soviet Union, Pakistan, Saudi Arabia, and the Gulf States).

There are a few criteria, however, against which Mamdani's and Abu-Lughod's valuable criticism has to be reassessed. Afghanistan is a strongly fragmented nation-state, mostly along ethnic and religious lines (Rubin 1995, Barfield 2010). This means that any traumatic event such as a military conflict has to be evaluated in terms of its impact on various ethno-religious communities that may have little in common with each other. Furthermore, the event itself, a continuous conflict now (in 2019) in its forty-first year, has been and still is also a fragmented phenomenon, which has affected different areas and populations of the country in different ways and to different degrees over the decades. The perception of it by such a diverse populace varies considerably. To consider just the post-2001 period, I will say that the Pashtun populations that live in the province where I carried out my fieldwork, Nangarhar, have been affected mostly by the struggle between a rising popular insurgency and Afghan military forces. US troops proper were engaged in the province to a very limited degree. What my informants experienced after 2001 has been mainly a brutal and internecine conflict among Afghans, with the participation of scores of foreign (mainly Pakistani) fighters. This is what they mainly had to cope with on a day-to-day basis, and what preoccupied their minds the most. The contribution of US troops has been mostly through stand-alone operations in remote rural areas of the province, and through the maintenance of a few Forward Operation Bases scattered in the territory. The overall perception that Pashtun people in Nangarhar had of the Americans involved in the conflict was mostly negative, being identified as interlopers and occupiers, as well as the main reason for the presence of an active insurgent movement in the area, which made their daily lives miserable and extremely dangerous. Those among my informants who felt resentment toward the presence of the US troops in Afghanistan, and criticized their own government for the support it gave to their occupation, did not necessarily support the Taliban insurgency as a result.

In other areas of the country, however, for example among Shi'a Hazara populations in Bamyan and Daykundi provinces, as well as in West Kabul, the perception of the international troops in the country was extremely positive, for a whole gamut of reasons that Melissa Chiovenda has thoroughly examined in her work (Chiovenda M. 2014, 2015). For Hazaras, then, the additional and renewed wave of violence that the post-2001 occupation has brought to the country has a completely different significance, and hence has contributed to very different ramifications both in the subjectivity of individual subjects as well as in the development of a particular group consciousness. For these reasons, the generic category "Afghans," or even "Afghanistan," when trying to assess the impact of a dramatic phenomenon such as the current or past conflict, is an inadequate unit of analysis. An intersectional approach that may break down diverse dimensions of identity present in the country's population is in this case preferable, in order to portray a more realistic picture.

Most of my Pashtun informants spoke of *jang*, "the war," when they wanted to reference the period of conflict between 1978 and the present. Multiple times, in my conversations with them, I had to ask an additional question to clarify to which "war" they were referring—the Soviet occupation, the civil war that followed it, the regime of the Taliban, or the current insurgency against the US-led coalition and US-backed Afghan government. Whatever the main "culprit" for these disparate strands of conflict, whatever the political meaning for any of them, in their perception these catastrophic wars were collapsed onto one, long, formless period of suffering and violence. My analysis of their personal and subjective relationship with, and negotiation of, an apparently seamless stream of violence, with its attendant deterioration of moral principles that, in their opinion, contributed to the perpetuation and increase of it, relies strongly on their "collapsed perception" of these long-lasting conflicts. That the idioms and performance of masculinity, and specifically of violence, had shifted in their public expression and implicit codification because of a never-ending "state of moral exception" since 1978 (Agamben 2005) was glaringly clear in the view of most among my informants. It was scarcely relevant for them, however, whether such perpetuation of the state of exception and its consequences was to be attributed to the Soviets, the Afghan warlords, the Taliban, or the US-led international troops, and what was its precise political genealogy. Habituation to, and routinization of, abuse, insecurity, social devastation, and unbearable life hardships had made it impossible for the average citizen to distinguish anymore between a "good" power broker and a "bad" power broker, of whatever sorts and dimensions, whether local or international. A generalized distrust of any political authority and widespread cynicism had rendered the "politicization of violence," which we, outside observers, are so used to underline, almost insignificant and pointless. What counts is the daily misery one is confronted by, and in pragmatic terms little counts as to who or what is the main culprit for it.

The shifts in meaning and performance of Pashtun idioms of masculinity that my informants felt had emerged during, and because of, the four decades of violent turmoil in the country, and which I will explore in this book, were very "real" for them, and had already been institutionalized (though to various degrees) into that "cultural baggage" that every one of them had been enculturated into and raised with. The dynamism one expects of any cultural tradition was marked in my informants' view by the perceived negative connotations that such dynamism had acquired for them in the wake of the conflict(s). Their subjectivities, the material with which I worked more closely, were severely impacted by such negative connotations, and part of their inner suffering derived precisely from this. From their point of view, thus, the politics behind the cultural material they had to cope with every day at a very personal level was of secondary importance, and from this standpoint I look at it in the following chapters.

2

Rohullah

Shifting Subjectivities and the Crafting
of a Private Masculinity

Prologue

Rohullah is one of the first people I met in Afghanistan, at the beginning of my
fieldwork, in the summer of 2009. His father, Asadullah, is a professor at Kabul
University, and has been for almost all his career—except a brief stint during the
Afghan civil war (1992–1996), when he moved to Pakistan. They are members
of the Mangal tribe (*qawm*, in Pashto), a group that has its traditional center in
the northeast corner of the province of Paktia. The grandfather of Asadullah
moved from the family's ancestral area to a village near the capital city of Paktia
province, Gardez (a couple of hours' drive southwest of Kabul), where they settled, and where most of the family is still living. Paktia province is a very traditional and conservative Pashtun area of Afghanistan, where people refer
proudly to the *qawmi laar* (loosely translated as "tribal system") as being still
a very strong aspect of social life. It also lacks a real cleavage between rural and
urban contexts: Gardez, the provincial capital, is little more than a square with
four roads departing from it, around which extends a big bazaar. People go to
Gardez from all over the province to buy and sell goods, meet friends, interact
with the state's administrative apparatus, and receive different types of services.
Gardez feels like just a large village: there is no university nor civilian airport, no
state-provided electricity (for most) nor running water. The sociocultural environment is extremely different from the one in Kabul. Cosmopolitanism, ethnic
diversity and personal "anonymity," educational infrastructures, and better
links to the outside world make Kabul a place where certain cultural norms find
a more "relaxed" application. In Paktia, the norms that define a "respectable
Pashtun" are still very much a feature of daily life. We will see that this contrast
has had a strong impact on Rohullah's inner trajectory.

Asadullah's appointment at Kabul University made him relocate early
from the village to Kabul, bringing the two daughters he already had and

Crafting Masculine Selves. Andrea Chiovenda, Oxford University Press (2020). © Oxford University Press.
DOI: 10.1093/oso/9780190073558.001.0001

his wife with him (circa 1978). Subsequently, one more daughter, three sons (Nur Mohammed, Zair, and Rohullah), two daughters, and one last son (Iqbal), in this order, were born in Kabul. Asadullah Mangal is one of the very few scholars of his generation who trained at the graduate level in a Western country, earning his PhD in 1977. When I reached Kabul for the first time in June 2009, he was more than happy to help me with my research. I was able to visit several times his village in Paktia province, where I stayed for a few days each time as a guest in his family's house. Zair and Nur Mohammed lived stably in the village back then—in 2012, Zair was forced to move to Kabul because of security problems linked to his job as a civil engineer in charge of the construction of roads in the provinces of Khost and Paktia.

The very first day that I met Asadullah Mangal in his office at Kabul University, in June 2009, Rohullah was with him, helping him solve some computer problems he was having. At that time, Rohullah was still a student at Kabul University; he has since graduated. During that first meeting, Rohullah remained in the background, respectful of the figure of his father, without speaking or addressing me. The importance of familial hierarchy is a feature of Pashtun society that I soon learned to recognize. Asadullah Mangal came to visit me a couple more times in the subsequent weeks, always accompanied by his son Rohullah. During these visits I introduced myself to Rohullah, and started working on building a relationship that has lasted ever since. His good knowledge of English, higher education, and an interest in self-introspection that he soon displayed gave him an important role in my research. We met regularly in Kabul in the following years, whenever I would drop by the capital. During my last period of fieldwork in Afghanistan, 2012–2013, Rohullah agreed to participate in my person-centered ethnographic project. We met one-on-one seven times, for a total of approximately twelve hours. He chose to conduct our interviews in English, which was useful practice for him in the first place, he told me. By then, however, I had already become fluent in Pashto, so I asked him often to integrate his English narration with the Pashto version of certain expressions or specific words that seemed more of interest to me.

The impact of a different social milieu

Rohullah Mangal was born in Kabul in (approximately) 1987. He lived in Kabul until 1992 with his parents and siblings. His paternal grandparents died before he could know either of them, while his maternal ones he remembers (they died in the recent past). His father talks about his own mother as being a very strong woman, someone who, in spite of being illiterate, inspired her children and guided them to an honorable way of life. Rohullah talks about his mother in similar terms, praising her fortitude and rectitude, and highlighting that, in spite of her being illiterate as well, she always pushed her children toward education and betterment in life.[1] Rohullah does not have clear memories of his early experiences in Kabul, except that his family had a comfortable life, in a well-furnished, well-equipped house, thanks to the efforts of his father and his academic position. Then the civil war started, in 1992, and Kabul became the center for internecine fighting. The city was heavily shelled by several different *mujaheddin*'s parties in conflict with one another, with ensuing chaos and civilian casualties. In order to protect his family from the dangers of the war, Asadullah Mangal moved everybody to his village near Gardez, in Paktia province, where one party among the *mujaheddin* had gained complete control of the area, and prevented the turmoil from spreading. Rohullah remembers how they left everything they owned in Kabul, and how they reached Gardez with almost nothing but their own personal effects. He remarked multiple times that the situation was extremely difficult for his family in the beginning. They had to rent a house in Gardez city, because none of their relatives would help them in getting by (Asadullah and his nuclear family never had a house of their own while living in the village, sharing their extended family's house with Asadullah's brothers and parents). He justified his relatives by saying that, in a situation such as that, everybody had to take care of their own families, and could not worry too much about others. We will see that relations with relatives (from the father's side) in the village will remain strained and conflictive. Rohullah's family remained in the rented house for two years, until he was about eight years old. Deprivation and struggle for survival are the most vivid memories he has of those first years after relocation. Interestingly, one of the examples he used to describe their problems was the presence of *jinn*s (i.e., spirits) in the rented house they occupied (which had the shape of a typical Pashtun house, with a main body; a courtyard for animals, a garden, and fruit trees; and a wall surrounding the whole complex). Since the very beginning, strange,

---

[1] In fact, Rohullah's three older sisters (who are older than all the male siblings) went to school and university, getting married in Kabul and having careers as physicians and schoolteachers. The two younger sisters, born during the war, could not attend school due to the security situation, and ended up being married off without attaining any academic degree.

inexplicable things happened in the house. A calf that was healthy during the evening was found dead the morning after. Several chickens died without explanation in front of Rohullah's mother. Noises at night were heard, as if something was hitting the walls of the house and doors. Rohullah told me that, having been raised in Kabul up to that point, he and his siblings did not know anything about jinns ("there are no jinns in Kabul—they usually live in areas where few people live, and in old abandoned houses. Our rented house had not been lived in for a long time, and was outside the city"). Yet his mother dreamed of the jinns, who urged her not to bother with them, pray, and keep on with their lives, because her children really needed her. It was not possible for me to ascertain whether Rohullah's mother dreamed first of the jinns, or experienced some uncanny events in the house prior to having the dreams. It may not make much difference anyway. The perception of the inexplicability of certain events may have been directly related to the highly stressful and uncertain situation that the mother was undergoing. A cultural background in which jinns are believed to be part of the "behavioral environment" (Hallowell 1955) will offer an available avenue to give shape and concreteness to the anguish and despair deriving from being uprooted and in dire straits, which in turn renders anguish and despair bearable to some degree. In fact, Rohullah himself spoke of jinns as existing entities, of which indeed the Qur'an talks, and which he has therefore no reason to doubt. Yet he approached the subject with balance and "rationality": "If you don't tease them, and bother them," he said to me, "they will not bother you." They apparently did kill the family's animals, though. For Rohullah, however, this was the patent, symbolic demonstration of the hopelessness of their condition at that time. Asadullah Mangal was seldom at home, while running a temporary business by transporting medicines from Kabul to Gardez, and selling them to the pharmacists, who had seen their stocks heavily reduced due to the war. The trip to Kabul and back was a dangerous one, and he managed to make ends meet for his family charging the pharmacists slightly more than the standard cost of the medicines. The resilience of his mother in the face of these hardships apparently strengthened the respect that Rohullah still displays of her, in spite of her being an illiterate woman.

During those first two years in Gardez, Rohullah's family subsisted through the small salary that Asadullah was providing, which paid for a cow and a few chickens and goats, from which they got milk, eggs, and meat on an irregular basis. Rohullah and the male children were put in school, while the female siblings had to stop attending. The older three females, who had already started university, or were finishing school in Kabul, remained there and continued. The younger two remained confined to their home in Gardez, due to the harsh norms enforced by the mujaheddin in the area. "They were like the Taliban," Rohullah commented, "they had a different name, but they behaved in the same way." In fact, despite the great fanfare with which the mujaheddin were publicized in the West during the anti-Soviet struggle, which still resonates somewhat today, their rule in the 1980s and early 1990s was largely characterized by a degree of religious fundamentalism similar to that of the Taliban regime a few years later. Girls were forbidden from public education, and women's life conditions worsened considerably. Not that the implementation of such norms found strong resistance in Paktia province. Rohullah remembers proudly how his mother insisted in sending her sons to school, while many of their neighbors considered public schools as venues for anti-Islamic propaganda that would eventually turn the students into *kafirs* (infidels).

The brothers of Asadullah who lived in the area, the main blood relatives of reference in a strongly patriarchal society, were reluctant to help his family, and, if anything, tried to take advantage of its position of weakness. Such weakness was due to the fact that all the male children that Asadullah had available at the time were still in their childhood or early adolescence. Nur Mohammed was fourteen years old, Zair twelve, and Rohullah only six (Iqbal was an infant). This made of Asadullah's family a "weak" family. Where numbers make power, and call for respect, Asadullah found himself in a position of vulnerability vis-à-vis his own brothers, who all had older sons, already able to bear arms (Asadullah's first three children had all been female). When it came the time to divide the land property that his father had left as inheritance, the brothers of Asadullah openly took the lion's share and left him with a smaller and less productive plot of land. They ridiculed his lack of adult offspring and, without even discussing, divided the land so to take the best sections for themselves. Rohullah, in recalling the facts, likely tapped from his father's memories as recounted to his sons at a later date—he would have probably been too young to remember. Nevertheless, the image of his father, emasculated by having to bow, if unwillingly, to the bullying of his brothers for the sake of pragmatism, sticks with Rohullah to this day, and is a counterpart to the subsequent events in the family and his life. The sense of humiliation, unfairness, and injustice that he derived from his father's treatment on the part of his brothers certainly influenced his subsequent behavior with his relatives and friends. While he did not have direct contact with his uncles (the brothers of his father), he did interact with their sons, his cousins. Of such relationships he remembers the continuous bullying he and his brothers had to go through. "They always made fun of us," he said; "they called us names and ridiculed us for being poor and not being able to stand up for our rights." The relationship between paternal cousins (*tarbur*, in Pashto) is traditionally fraught with jealousy and competition, and the term *tarburwali*, literally "the way of being cousins," has become a synonym for enmity, or *dukhmani* in Pashto. In fact, when someone wants to refer to a paternal cousin in a way that does not imply enmity, he uses the term for the maternal cousin, *de tror zoy*.[2] Rohullah remembers that, despite what Asadullah had to go through with his brothers (or maybe because of it), he always encouraged Rohullah and his older brothers not to be weak, to keep a strong attitude (*kamzur ma wsega*), although, at the same time, he constantly deplored the perpetuation of relationships of *dukhmani* and feuding. The discussion of the concept and relations of power and respect that Rohullah gave me is interesting in this regard. He explained:

> In Paktia, if you don't have power [qowat] nobody respects you. They will always take advantage of you and abuse your rights. Power is given by numbers, by the number of sons and brothers that you can gather in protection of what is yours. My father at that time was alone. We were too young, and he had to confront his own brothers, who all had much older children than us. If you don't have numbers in your family people won't treat you with respect, they will consider you a weak person, who is unable to protect his rights. They will not be afraid of you.

ANDREA—*So, respect actually comes from fear?*

ROHULLAH—*Yes, people will treat you well because they are afraid of what you will be able to do against them if they disrespect you.*

ANDREA—*Is there anything else that brings respect to a person?*

---

[2] I have mentioned in the previous chapter the importance of the patrilineal system in Pashtun society, and how it is central to shaping certain interpersonal relations. It is true that "real," full lineage mates are considered only the blood relatives on the male side, and that they compete for the resources that the family will allocate to its members (like in this case, the father's inheritance). Yet the relatives one has on his or her mother's side, precisely because they are not in a competitive relationship with ego, are often considered to be the "good" ones, those with whom one can have a more close, frank, and sincere relationship. Though they might not have a say in the most economically and politically important decisions that a family will have to make, still relatives on the mother's side play an important role at the individual level to ensure balanced familial interactions and reduce frictions where possible.

ROHULLAH—*Well, obviously, if you are a learned person, or if you behave like a pious Muslim, or if you display wisdom and knowledge of your people's laar* [customs], *they will think highly of you, but they still will consider you weak, and someone sooner or later will come and take what is yours. Also money brings respect. A rich person will be treated better than a poor person. They considered us less because we had no money.*

Thus, Asadullah was doubly "weak": he was without adult male offspring that could defend his rights, and was not well-to-do enough to be considered fully respectable. Such situation continued for the first year and half of the family's stay in Gardez. Subsequently, Asadullah found a job in an international NGO that worked with refugees in Pakistan, and started bringing more money home. He would reside in Peshawar, Pakistan, most of the time, and come home to Gardez for a couple of days every month. The family saved up a good amount of money and started building a new spacious house in the village of his father (a *qala*), where they eventually went to live. Rohullah says:

Then we became rich. We saved money and built a house for ourselves in my father's village. Things started to change, and we were given more respect by the people in the village. I remember that when I was maybe twelve, I was playing with my cousins and other kids in a field. We were playing a game in which we had to hit a ball with a stick. One of my cousins [one of his father's brothers' sons] *was still teasing me, he was calling me bad names. I was a handsome kid at that time, and he was calling me "bacha." Then, all the times that in the past he and his brothers had teased us, and disrespected us, all those memories came back to my mind. I had grown up, and I did not want to go through that any longer. I grabbed my stick hard, approached him, and started beating him with the stick. He was older than me, maybe nineteen or twenty. But I was big too by then. I broke his arm. He went home and did not say anything to his father. He said that a cow had broken his arm when it kicked him. He did not want to be ridiculed for having been beaten up by his younger cousin.*

Rohullah's cousin called him "bacha" ("boy" in Persian), a term with which people describe the male children who are sometimes used as sexual toys by powerful male adults in the community. The practice brings shame to the child and his family, although it is somehow seen as a display of personal power by the adults who engage in it. Rohullah says he was a "handsome" boy, and remembers that he had to be escorted by his older brothers when he went out of the house away from the beaten path, in order not to fall prey to those who might have wanted to take advantage of him, for whatever reason it might be (it has been reported to me that sometimes the abuse of children is used as a weapon during inter-family feuds). Even at his young age of twelve, Rohullah was already well aware of the insulting attempt at feminization that the word brought about, and reacted with rage. The sentiment of offense to one's masculine honor in this case was perhaps magnified by the memory of the years of abuses that he and his family had to go through in the past, and from the subdued, emasculated image of his father that he was probably trying to rebel to. Both his older age and, probably, the realization that through their jump up in the social ladder his family had acquired a stronger public validation, rendered him bold enough to stand up to his abusive cousins, and finally heed his father's advice of "not being weak" (*kamzur ma wsiga*).

Crafting an alternative masculinity

In fact, age twelve is when Rohullah remembers to have undergone a fundamental personal change, as a conscious choice. From the time he moved to Gardez, until he was approximately twelve years old, Rohullah reports to have been a very quiet child. He did not get into fights, nor did he act aggressively. He did not respond to provocations and shied away from situations that could have turned violent. This in spite of the hostile environment that surrounded him in the elementary school he attended.

My classmates did not like me. They saw me as the city boy who comes to the countryside, and didn't like it. I had different manners, a different way of dressing, a different way of speaking [Pashto]. They would come to class with their farming clothes, with which they would work in the fields after school, while I had always clean clothes that came from the city [Kabul]—I did not have to work in the fields. They were jealous [bakhil] because I was more intelligent, I was better educated, and the teachers liked me. I got better grades than they got. They would tease me and make fun of me. When we played volleyball in the school yard they used to hit me with the ball intentionally, just for fun, and laughed. My school was located one and half hours away from my home. I had to walk that much time to get there every morning.

Sometimes, on my way to school or back to my house, I was bullied by some schoolmates. They made fun of me, threw rocks at me for fun, sometimes roughed me up [haghuay za wahalam]. *They had this aggressive attitude that I did not have. They picked fights between each other. I did not participate, because I did not want to fight. Also, I liked going to school, my parents taught me that it was a good thing for me to go to school. The other kids did not like it. Their families sent them to school just because they needed to read and write in order to get a decent job, but in fact they thought that education was something that would make you less Islamic, not a good Muslim. They were ignorant, ignorant* [besauaday, jahili].

ANDREA—*Why do you think they had this aggressive attitude?*

ROHULLAH—*I think it had a lot to do with the war. They were born and raised in this provincial environment where the war against the Russians, and then the civil war after that, affected society a lot. There was a lot of violence every day, a lot of conflicts also among local people, everybody was against everybody else. It's difficult to live that way. After a while I guess it becomes normal. But it also has to do with* ghairat. *Ghairat is deeply engrained in Pashtun culture. A Pashtun man has to be* ghairati [see below].

Rohullah is experiencing personally the traumas of children born and raised in a war zone—his classmates. He becomes the scapegoat for a set of moral values and ethical comportments that have already changed and adjusted to the unforgiving necessities of the long-lasting conflict. He is keenly aware, and with remarkable perceptivity, of the psychological consequences that the widespread and institutionalized violence had had on his peers. As many of my middle-aged informants have repeatedly stated, morals and ethics in the Pashtun context received a painful blow during the war. The crucial concept of ghairat (manly prowess, courage, fearlessness), which had traditionally been linked to the honor of the family and the individual, in turn indissolubly tied to strictly regulated gender relations between men and women, and the control by men over women's autonomy in the public sphere, during the many years of war had shifted in meaning, more and more becoming coterminous with sheer power, violence, and self-aggrandizement, due to the routinization and institutionalization of a moral state of exception introduced by the conflict (see Agamben 2005, and below for details). In a feedback effect (reminiscent of Fredrik Barth's transactional dynamics, Barth 1966), those aberrant behaviors, which would be normally stigmatized by society, and that had been at first justified by the equally abnormal necessities of a gruesome war, had been over time institutionalized, and had in turn created new moral values, and attendant ethical behaviors, which, at the time of Rohullah's early adolescence, were being already internalized by his classmates. Raised mostly in an urban environment, and enculturated into more traditional Pashtun honor-related values by his father (*kabarjan ma wsa, liken kamzur ma wsega*), Rohullah finds himself at a loss when he clashes against a rural Pashtun context scarred by the war, and that already presents values and ethics adapted to the now-permanent state of exception.

Little by little, however, he grew fed up with such situation of helplessness and victimization to which his mild-mannered demeanor was condemning him. Some time in sixth grade, he decided to turn things around.

> *I did not want to be considered weak any longer. The fact that I was quiet, shy, and did not want to end up in fights was mistaken for weakness. I wanted that to stop. My father had raised me stressing values of peacefulness and calmness, unlike the fathers of the other kids did, and these values were what I considered right. But I was having a hard life outside the house, it was a nightmare . . . and I wanted it to stop. I chose to learn how to be more aggressive and assertive. I started following more closely what my elders and the stronger men in the village were doing, how they were behaving. I started attending the* marakas *and the* jirgas *in the village* [councils of elders to discuss and solve intracommunity conflicts], *so that I could see how they were asserting themselves and fighting for their rights. I also started to respond to the other kids' provocations, and to fight back. I fought more often over time.*

ANDREA—*How did this new behavior of yours make you feel? How did you feel about fighting?*

ROHULLAH—*I did not like it. I felt there was something wrong about it, and I remember that when I got home after a brawl* [lanja] *I felt unhappy* [khapa]. *But I had to do it. It was about survival. I had to survive in that world, I could not just let things go like they had gone until that time. I had to survive a bad situation. I used to tell myself that it was not my fault, I had to do it.*

ANDREA—*Was it like this all the times you came back home after a fight?*

ROHULLAH—*Yes, but after some time something changed as well. You see, a Pashtun man has to be ghairati, he has to have ghairat. It is something very important for a Pashtun man. This is important for me as well. If need be* [ka cherta da pa kaar wi], *you have to show that you are a ghairati man. You have to protect your rights. When you do so, people around you admire you, and give you respect. The more you show ghairat, the more people talk about you in a respectful way. You start to have a reputation* [nuum], *you start to have power* [qowat]. *Until I kept quiet and did not fight, I had very few friends. After I started fighting back and standing up to the bullies who bothered me, a lot of other kids started hanging out with me, looked for my company. I felt I had gained power. It made me feel good. After some time I think I became like addicted* [amali] *to the power that my aggressive behavior was giving me. I enjoyed a lot of popularity among my schoolmates. They wanted to play with me.*

ANDREA—*You did not feel bad any longer for fighting with other kids?*

ROHULLAH—*Well, yes, I still felt bad about it, I still felt I should not have behaved like that. My father would not have liked it. But it was a strange feeling, it gave me pleasure to have all those kids following me as if I was their leader. I would also feel pleasure when I told my friends the stories of my brawls and fights. They were all excited and in admiration of me. So I think I really got out of control, and started behaving more and more violently and aggressively. Too much. You know, there are two kinds of ghairat. There is good ghairat, and bad ghairat. For example, imagine you are walking in the street with your sister, and a strange man passes by and looks at your sister intensely, as if she knew her. If you start speculating that they are having an illicit relationship, and kill them both without even inquiring with any of them, you are not a ghairati man, you are ignorant and stupid* [besauada aw kamaqal]. *On the contrary, if you discover that your wife has been the object of bad verbal or physical harassment against her will, and you kill the culprit, this is good ghairat, something obligatory* [majbur]. *However, this being ghairati, it also brings lust for power* [ghoror], *as it has happened for me. It makes you feel good, important. And also, you have to continuously demonstrate you are ghairati to all the other people, so you try to always top the action that you have undertaken the previous times, to impress others, and fulfill expectations. Violence escalates in this way. But I also think that greed for power is something inherent in human nature. Everybody is somehow affected by it. If a structure like the state does not exist, or is weak, like here* [in Afghanistan], *there will be no restraint in people. There should be a power like the state to punish people and prevent them from becoming so violent.*

The intolerable frustration that Rohullah is subject to for years during his childhood reaches a peak in his early adolescence, and something snaps. Or rather, I argue, something coalesces around an alternative pole of self-representation. The years he spent as a spectator and victim of the dramatically different moral and ethical context that he found upon moving to the village now formed a critical mass of unconscious experience from which to tap in order to save himself from a dramatic personal and social situation. I would argue that such critical mass likely constituted what Bion calls "beta-elements" (Bion 1962)—an indistinct cacophony of sensory material (emotions) that needs to be processed in order to emerge as "alpha-elements," that is, material that can be thought and "wake-dreamed." The years spent until then in the village represented no doubt a painful and arresting experience for Rohullah, who nonetheless managed unconsciously to operate on the cacophony of his beta-elements successfully enough to trigger a healthy alpha-function, which gave new meaning to the beta-elements and processed them into alpha-elements, accessible to and "manipulable" by consciousness. (For the importance of meaning-creation within any psychic process, see Bruner 1990.) He remembers how the elders and his older brothers came to embody slowly a new paradigm that he step-by-step embraced. He internalized over time the same values and ethics that he was the victim of, and unconsciously opened a new avenue for expressing himself in a culturally meaningful and accepted way, a way that would assure his "social survival," as he underlines. This, I argue, is an index of meaning-creation. Rohullah rendered meaningful for himself, if only in terms of "survival," those moral principles that he observed turned into pragmatic behavior by those who surrounded him. He actively endorsed the constitutive elements of a hegemonic masculinity (Connell 1987, 1995), as defined in the context of Pashtun rural life, rejecting at the same time those that indexed his own subordinate masculinity, and which had condemned him up to that point to social ostracism. Such strategy for protecting social survival is adopted mostly in a non-conscious fashion, I believe. Rohullah does remember clearly when he decided to shift gears, and to start behaving in a more locally appropriate way. On the other hand, he seems to have been unaware of the laborious and difficult work, taking place backstage, needed to internalize those norms that were hurting him, in order to turn them to his own advantage. The pain and anguish he went through before he was ready to adjust his behavior was the backdrop of the unconscious labor he was putting up to craft a different masculine self, which could hold against the attacks his original masculine self was succumbing to. What I am tentatively describing here is the crafting of diverse, interconnected, yet dissociated states of subjectivity (or, in the words of Philip Bromberg, dissociated states of consciousness. Bromberg 1996: 267–290), which index the presence, coexistence, and potential conflict of multiple selves. By dissociated states of subjectivity I do not mean a pathological cognitive disorder.[3] Rather, I mean the constructive and psychically healthy capability of accessing different sets of meanings and identifications, which serve different purposes at different stages of one's life trajectory, and which are kept in a fruitful, if often painful, dialogue with each other. From such dialogue emerges the "illusion" of the unicity and continuity of the self. Katherine Pratt Ewing, in fact, wrote that "The experience of a cohesive self must not simply be dismissed or ignored simply because it is illusory" (Ewing 1990: 263), while psychologist Roger Frie termed it a "necessary illusion" (Frie 2008a). In the same article, though working from within an ego-psychological theoretical framework, Ewing proposed already a similar paradigm to the one I elaborate here for understanding inner dynamics, which she defined as "shifting selves" (Ewing 1990, one of several truly pioneering articles that Ewing produced in the same period. Ewing 1987, 1991, 1992). Interpersonal/relational psychoanalysis, particularly with the work of Philip Bromberg, has contextualized the fluidity of subjective experience within a more comprehensive theory of mind, in which dissociation, both as a positive and negative process, becomes a central feature of psychic life.[4] From this standpoint, the unconscious should be also seen as the locus for "the suspension or deterioration of linkages between self-states, preventing certain aspects of the self—along with their respective constellations of affects, memories, values, and cognitive capacities—from achieving access to the personality within the same state of consciousness" (Bromberg 1993: 182). In this sense, when the access to one or more of such self-states, or sets of meanings, is for any reason unconsciously impeded, and barred from the fruitful dialogue between each other, only then I believe may we detect the dysfunctional use of such dissociated states of subjectivity,

---

[3] In spite of the scant attention that they received in classical Freudian and Kleinian psychoanalysis, dissociative processes have been acknowledged in contemporary psychoanalytic theory to be, under certain circumstances, potentially fruitful and positive mechanisms. See, in this regard, the works of Philip Bromberg (2003), Peter Goldberg (1995), and Alan Roland (2011).

[4] Let us not forget that the American psychiatrist Harry Stack Sullivan, who is credited to have spearheaded the interpersonal school of psychoanalysis, and who worked closely and fruitfully with anthropologist Edward Sapir, noted that, "For all I know, every human being has as many personalities as he has interpersonal relations" (Sullivan 1950: 221).

and eventually that phenomenon that in classical psychoanalytic literature is called repression.

In Rohullah's case, his account of those difficult adolescent years in the village shows, in my opinion, how he usefully not only managed to craft an alternative set of subjective meanings (an alternative self), better adjusted to the life in the village, but, more important, how he did not eliminate from the dialogue with his alternative self the self that he embodied when he first arrived in the village—the quiet, shy, and studious boy. When Rohullah finally decided to shift gears and proactively embrace a subjectivity better adjusted to the village environment, he also shifted from a condition of *dissociation*, in which he had barred out the "unformulated experience" (Stern D. B. 2010) of the self he was unconsciously "preparing" to embody in the village, to a condition of *conflict*, in which both sets of meanings and subjective states became accessible, "awaken" to each other, and inhabitable, though conflictually (see Bromberg 2003: 702). In other words, he realized that he had already construed for himself an alternative self that could be endorsed and enacted to save him socially, and which he now could perceive in contrast with the previous urban self, "learned" through family relationships. The emergence of the conflict, which replaced the previous deep dissociation, maintained the pain that his condition entailed, but pulled out Rohullah from the static condition of paralysis that characterized his first years in the village. Until this existential breakthrough and change in behavior took hold, Rohullah was "stuck" in one self-state, without the capability of accessing other self-states (subjectivities) that could have been pragmatically better attuned to the changed reality and environment of the village, and which he had already been shaping, unaware, in a severed and dissociated form. About this specific condition, Philip Bromberg writes that:

> *A centrally defining hallmark of such* [an individual] *would be the dominance of a concrete state of mind in which the experience of internal conflict is only remote and briefly possible (if at all). . . . As a result of the subjective isolation of discretely organized realms of self-experience, data that are incompatible with the ongoing self-state are denied simultaneous access to consciousness. . . . The experience of conflict is structurally possible but psychodynamically avoided.* (Bromberg 1993: 182–183)

Rohullah feels the burning contradiction and incoherence between what he had to become for his survival's sake, and what he had been taught to be by his father and mother in their urban environment. These two sets of meanings painfully see each other, they are aware of the presence of each other, and he is, if unconsciously, aware of the necessity that one take precedence over the other under such contingencies. As Bromberg would put it, Rohullah now is *more conflicted than dissociated*.

And one subjective state did take precedence, imperiously at that. How to interpret the unexpected inebriation with power, force, and violence that Rohullah reports to have experienced? To begin with, I think it is important to bear in mind that Rohullah never bars out from some degree of awareness the self with which he had come to the village—the shy boy who shuns violence. The use and abuse of power and force that he carries out gives him pleasurable feelings, but a very bitter aftertaste. The interconnection between his alternative selves (his states of subjectivity) is constantly operating. His words describe "becoming addicted" to the new condition of power and popularity he attains among his peers, the "pleasure" that he feels at being finally not only accepted by his peers, but unexpectedly put on the pedestal as a model for others. There is certainly a degree of narcissistic seduction and fulfillment in this dynamic. Yet, in the young Rohullah who goes overboard with his new self, I perceive the adolescent, desperate attempt at being recognized, validated, and appreciated by his peers, and social environment as a whole. It seems as though those many years of rejection and suppression by his social environment had created the conditions for an uncontrollable explosion of emotions, the inebriation with a commonality and "togetherness" with his peers that he does not want to put a limit to. At the same time, it is noticeable that Rohullah does not react with hate or rejection to the sadistic treatment reserved for him by the young boys in the village. On the contrary, he strongly identifies with the different cultural context he finds there, he wants to move from the subordinated masculinity position he was in when he went to the village, to that of hegemonic masculinity recognized by his peers, and embarks on a quest to win back those who "hate" him. There is no masochistic acceptance of a subjugated positionality. This, I believe, speaks as much of the psychic dynamics of Rohullah as of the power and all-encompassing nature of sociocultural arrangements in a Pashtun traditional rural context such as Paktia province. The connective selves (Joseph 1999: 2–17), or rather, the connectivity between the subjectivity of people who live interdependently of one another, in a face-to-face environment, operates strongly, as we can see, and still, it does not erase the individuality and "individuation" of each one of them.

The effort at re-establishing an internal balance

Rohullah's life in the village in Paktia continued to sail amid problems and turmoil. The absence of his father meant, among other things, that there was no one available to farm the land that had been assigned to him upon the death of his father. After a few years of cultivating Asadullah's plot for themselves, his brothers refused to give it back to Asadullah's family, who in the meantime had saved enough money to pay for hired farmers to cultivate it. Tempers flared up in the family, and Rohullah remembers that they had to physically fight multiple times against his cousins and uncles in order to "convince" them to leave the plot of land to them, which eventually they did.

Nur Mohammed [his oldest brother, born circa 1979] *was not inclined to fighting. He was the one who had responsibility of the household during all those years that my father had to spend away from home. He had to learn to be level-headed and wise, and could not afford to be just as a hothead as me and Zair were* [Zair is the second oldest son, born circa 1981]. *Zair did not have any specific tasks or responsibility, and he also showed to be much more naturally* [pa khpal] *aggressive than we were. He used to follow the big guys and the elders in the community, to learn from them as to how to behave. When I decided to turn around things for myself, and become more assertive, I looked up to him as an example. You know, Zair spent a lot of time in the village when he was a child, he knew better than I did how to comport himself among aggressive people. But also, he liked to be like that. He was more like them than I was. After what happened with the brothers of my father, my father decided that Zair and I would be the ones to uphold and defend the family's rights. We were old enough by then. Zair was a powerful guy, he was respected. Sometimes he also behaved abusively, though* [zalim]. *Yet, the environment in Paktia is conducive to this, they make you behave like this. He was the right guy to do it. There, acting aggressively becomes an obligation* [majburiat]. *If you don't do so, people won't respect you.*

In the absence of his father, Rohullah chooses his next oldest brother as a paradigm to follow (Zair), the one person who at that moment embodies the archetype of the "real" Pashtun man that Rohullah strives to become at this point in his life, the man who bears the banner of the hegemonic masculinity culturally shared and accepted in the rural, traditional context in which he is living. Rohullah in fact continued in high school to be an assertive person, and to gather followers, like a sort of gang leader. His father was far away, and could not do much to rein him in. Yet, one day, in one of his fights, Rohullah finally came across the wrong person.

I had become a troublemaker [badmash]. *One day, when I was in tenth grade, I had a bad fight with the son of a* kumandan [a militia commander]. *We both got injured in the fight, and went home to our families after swearing to take* badal [revenge] *against each other. A few days later, my father came back from Peshawar to bring the salary home, and he was informed about the incident. He ordered me right away to leave for Peshawar with him. I finished the tenth and eleventh grade in Peshawar. My school there was a very good school. My classmates were smart and studied a lot. They were not like my classmates in the village, they did not fight and quarrel all the time. I could just think of my study, without worrying about much else. I could imitate and emulate my father. He solved problems by talking, not fighting. I felt relieved, I saw that I could actually live in a different way. I said to myself, I want to be like these guys, not like I was in the village. I never fought even once when I was in Peshawar. I did my twelfth grade in Kabul. In fact, after graduating from high school I went back to the village, and I thought: this is not me, I don't want to live like this. A few weeks later I left to Kabul, where I enrolled in university.*

Peshawar, a large, cosmopolitan, intellectual, and then-peaceful city, constitutes a sharp change in Rohullah's life. Not only does he join his father again in everyday life, but he is catapulted back into an environment in which the state of subjectivity that he forced upon himself in order to adjust to life in the village (his alternative self, that is) is not necessary any longer, and may even possibly be maladaptive. The line of communication that he managed to keep constantly open between the self and subjectivity with which he arrived in the village, and those which he trained himself into during his life there, proves now functional to his new readjustment. Not having completely barred out from consciousness the subjective state in which he was when he reached the village at first, allows him now, in Peshawar, to reconnect stably and positively with it, with his "previous" self. A never completely interrupted dialogue gives him the chance to not only fully recuperate pieces of his subjectivity/self that he had stored away, but to choose rather consciously to endow them with the pre-eminence he had previously denied them. In Peshawar, Rohullah finds a multivocal set of hegemonic *masculinities* (see Demetriou 2001), which compete shoulder-to-shoulder within a more open

moral landscape. He is not forced any longer to (unconsciously) choose between a crippling subordinate masculine condition and a socially rewarding
hegemonic one. He is now able to align his own way of being a respectable
Pashtun man to one of the several ideal types of masculinity that in Peshawar
are equally culturally acceptable, and for this equally hegemonic.

And yet, although the school experience in Peshawar revitalized these
aspects of Rohullah's original self, still it was difficult to completely obliterate
the kind of person that he had "learned" to be in the village.

> One time in Kabul, I was in my last year of high school, when Karzai was
> president for the second year [2003]. I always sat in the front seats, because
> I was smart, I knew everything the teacher was talking about. . . . The other
> students were lazy. . . . One morning I found one of my classmates, a Hazara
> guy [one of the ethnic minorities of Afghanistan], sitting in my seat. I told
> him to move, but he refused to move. I asked him again, and he did not move.
> So I punched him in the face, and, without saying anything, he got up and left
> my seat. For a couple of months afterwards I was afraid that he would gather
> his friends and relatives and come to school with them to take revenge for
> what had happened. But nothing happened. Kabul is different from my village
> in Paktia.[5]

The first months in university, in Kabul, were no different in this regard.
During a Pashto language exam, he got into a fight with another student, right
in the classroom where the exam was being held. The friends of both rushed
to help, and the scuffle turned into a big brawl. Rohullah had to have his father intervene with the dean of the university in order not to be expelled.
Other participants in the brawl were expelled. Rohullah commented on both
events briefly, with a sardonic smile on his face: "Yeah, I was still a little bit of a
badmash. From time to time I still now end up in trouble with aggressiveness

[5] In analyzing this incident it is useful to consider also the ethnic component. Pashtuns have historically been the dominant ethnicity in Afghanistan, both from the political and social point of view.
Conversely, Hazaras have been the downtrodden, subject to decades of discrimination and persecution because of their Shi'a faith and supposed Mongol ancestry. Only after the demise of the Taliban,
under the protection of the international military forces, Hazaras have managed to win for themselves new political and social opportunities. To this day, however, many Pashtuns express (either
publicly or privately) contempt, acrimony, and resentment toward Hazaras (all the more because
of their newly acquired visibility). Such attitude might have been present also in Rohullah at the
moment of his act of bullying against his Hazara classmate. Melissa Chiovenda and I analyzed the
psychological nuances of this ongoing, though inconspicuous, interethnic conflict in a previous publication (Chiovenda A. and Chiovenda M. 2018).

and violence. But only when it is really necessary." My countertransferential reaction to this narrative was the marked perception that, in telling me the stories about his fights, particularly the ones happened more recently, he was feeling the same narcissistic pleasure that he used to feel when he recounted his "heroic" deeds to his friends in the village. He was, if unconsciously, trying to gain that same social capital with me as he looked for in the village, when he bragged about his virile accomplishments with his friends. In Kabul, with the kind of life that he is conducting there, he does not need any longer to display himself in such "hypermasculine" way. Clearly, though, a piece of his "village self" was still firmly with him. As a matter of fact, Rohullah does not deny that even now, after all these years, the self-inflicted "behavioral training" he went through in the village informs his daily life, albeit only to a certain degree. Put in another way, the strong, and somewhat traumatic, process of enculturation that he underwent after he moved from Kabul to Paktia, as well as the social constraints that he had to suffer there (which forced him to change his demeanor for the sake of "survival"), clearly affected a very deep layer of Rohullah's subjectivity as a whole. The line of communication between his states of consciousness, in Bromberg's parlance, or states of subjectivity, as I would rather call them, is still open, although now it works with an opposite flow, and with a reversed balance.

He gave me an example of this phenomenon, and the meaning for him of concepts such as honor (*izzat*) and determination/courage (ghairat), by describing what had happened to him just one day prior to one of our last interviews, in May 2013. We had the interview on a Sunday, and one day earlier, on the Saturday, he had had a fight with an in-law at his wedding. This in-law, Ahmed, had gone to Rohullah's wedding in 2009, four years earlier. Rohullah's wedding party (*mrasem*) was attended by hundreds of people from both families, who had come to Kabul on purpose from the two Pashtun provinces where the families sprang from, Paktia and Logar. Both relatives and friends of repute were present, including elders, army officers, political figures, and community leaders. At the end of the party, Ahmed, without an apparent reason (which is to be demonstrated), started making trouble, yelling and being disrespectful to other guests. Rohullah warned him a couple of times to stop it, but he continued. Finally, Ahmed and Rohullah started fighting, which turned quickly in a general brawl. Both ended up shaken, and Ahmed's side claimed to have "won" the contest, to have had the upper hand in the end. As it happens, Rohullah did not forget about it. His honor (*izzat*) had been compromised in front of a lot of people, for two reasons. First, because Ahmed had disrespected Rohullah's guests, and his wedding as a whole, causing embarrassment and shame (*sharm*). Second, because Ahmed walked away bragging to have "won" the fight. Ever since the wedding, there had been a stain on Rohullah's and his family's honor, that needed to be washed away. Positionally, Rohullah had been since then in a condition of diminished public respectability (because of his compromised izzat). Performatively, Rohullah had been ever since expected to demonstrate practically the willingness and courage to wash away the stain (i.e., being a ghairati man). He had not been considered *beghairata* (without ghairat) until that Saturday, because he had always publicly declared the willingness to redress the wrong suffered, to "take revenge" (*badal*). If, in the beginning, he had said, "That's ok, I don't care, just let it go," he would have been considered beghairata right from that moment, after the incident. Nevertheless, people had been waiting for him to act, to do something about it, to demonstrate in practice that he had ghairat. Finally, he was invited to Ahmed's wedding that Saturday, as expected. He went there, after having prepared a course of action with nine of his friends beforehand. He had called his older brother Zair, to alert him about what was going to happen, and to have some advice. Zair, in spite of his past as the "tough guy" in the village in Paktia, and despite having been Rohullah's role model in situations like this one, urged him not to proceed with his plan, and to let go. "Just leave it, let it go," Zair told Rohullah. "We have bigger problems now that we have to solve, as a family. Don't get involved in one more trouble that will last for a long time." Yet Rohullah disregarded Zair's advice. Taking revenge for Ahmed's disrespectful conduct, and performing appropriately for his audience (i.e., everybody else, including Zair), was too important at this point in his life. Right when the bride and Ahmed were about to leave the party in their car, Rohullah and his friends surrounded Ahmed and started beating him. Many other people joined in, obviously, and the police had to intervene, arresting three from Rohullah's side (including him), and two from Ahmed's side. Incidentally, the police was already present precisely because it had been alerted about the possible disturbance by Ahmed himself, who knew that Rohullah had not forgotten about what had happened at his wedding. The five men were released upon signing a paper wherein they promised that the "feud" would end there. In reality, the feud will not likely end so easily, in Rohullah's opinion. This is because Rohullah and his friends took care to cause a bigger brawl than Ahmed and his friends had at Rohullah's wedding, and, especially, because the bride was inadvertently slapped in the face during the fighting, and fell to the ground—which is considered a very injurious thing. Yet Rohullah felt that he had to "outdo" Ahmed in creating a disturbance, if he really wanted to wash away the dishonor of Ahmed bragging about winning the fight four years earlier at Rohullah's wedding. There were two reasons why Rohullah's honor had been compromised, and both had to be taken care of. So, now Rohullah is relieved: he is a ghairati man in front of everybody, and his family's honor is restored to its previous condition. At least until Ahmed finds a way to outdo Rohullah in return, and so on, and so on.

By presenting to me this insight into his recent daily life, Rohullah was communicating to me how much he had internalized and rendered meaningful for himself those principles, moral tenets, and ethical behaviors that he had to learn, if unwittingly, in order to (socially) survive in the village in Paktia (i.e., the patterns of hegemonic masculinity predominating in Paktia). At the end of our interview, he added:

> I did what I did at Ahmed's wedding because I had to. It was an obligation [majburiat]. I did not enjoy what I did. But I had to do it. Izzat and ghairat are important here, among us. People talk about you, they all give you a hard time. If I was living in the United States, I would have let it go. I would not have cared. But here, in Afghanistan, if I had let it go, my kids one day would have asked me: "Why did you not punish Ahmed for what he did to you, dad?" I am proud that I can show everybody that I am a respectable person.

In the same breath, Rohullah says, "I did not enjoy what I did" and "I am proud that I can show everybody that I am a respectable person." Yet the contradiction in Rohullah's narrative is only illusory. I argue that the Rohullah that emerged from the village in a new shape was no less a legitimate Rohullah than the original Rohullah, the child who was bullied by his peers until he was twelve years old, who did not want to engage in violence, and who continued to feel unhappy after each brawl he got into. He feels the disconcerting symptoms, however, of the conflict between these contrasting selves that he alternatively, and unconsciously, embraces under different circumstances in his life. The dissociated, yet present and engaged in an unconscious communication, state of subjectivity that makes Rohullah feel uneasy and uncomfortable when he behaves like an aggressive and vindictive person is counterbalanced by the state of subjectivity (his alternative self) that he developed while he lived in the village, in order to preserve his psychological balance, his social survival, and to ensure the validation by his social environment. The latter prescribes certain cultural standards of behavior in order to be considered a socially appropriate masculine individual (i.e., patterns of hegemonic masculinity). Such standards have been pragmatically endorsed by Rohullah, who has internalized them and legitimized them to himself. Under certain circumstances, in certain contingencies, they represent an integral (and "true") part of his subjectivity as much as does the moral abhorrence that he felt as a child, and still feels now, at the necessity to use violence in order to (socially) survive.

Rohullah has managed to navigate successfully enough of these seemingly contradictory aspects of his subjectivity (these two selves, as it were). He has managed to modulate the weight that, within a coherent sense of self, each of these aspects has to hold. Indeed, he now has permanently moved to Kabul, started a family there, and has no intention of going back to the village, where his brother Nur Mohammed and his family (and, until very recently, Zair and his family) are still living. In a powerful passage of one of our previous interviews, he confessed:

> I hate when I have to go to the village. I try to go as little as possible, and when I go, I stay for just a few days. Everybody is edgy there, everybody is always anxious, always ready for something bad to happen. They are aggressive, pushy. It makes me uncomfortable. When I go there, I change, I become a different person. I start doing things like they do them. I start to talk like them, to act like them. I also become more aggressive. I have to adjust to how they do things, because otherwise they would consider me a weak person, they would make fun of me. I know the rules, I know how to behave, I adjust. Every time I get in the car to go back to Kabul from the village, I feel like a weight has been lifted off my shoulders.

Thus, although Rohullah has certainly internalized and incorporated within his private sets of meanings the values and ethics of the life in the village, and has to some degree brought them back unconsciously with himself to a different life in Kabul, the village environment now presents for him a degree of intensity that he feels as unacceptable. When he is in the village, the facets of the selves that emerged from our conversations are not in harmony anymore. As a symptom of his unconscious processes, he perceives a disturbing unbalance in favor of one of them, the one he rationally disavows more, but that he feels he needs more to "survive" in the village. To reiterate: the self he embodied when he first moved to the village from Kabul, the quiet and shy child who wanted nothing to do with violence, was so maladjusted to life in the village that he rejected it after a few years of suffering. Social pressures and cultural incompatibility made the self-configuration he held upon arriving in the village unsuitable to appropriate social life in that context. In that particular life contingency, he disavowed that original self, those states of subjectivity, and worked hard toward building "alternative" ones, which would help him integrate in the new environment. Yet, although this "alternative" self was born out of pragmatism and the need for social adjustment, it has nonetheless become an integral part of his being, an aspect of the "authentic" Rohullah, a legitimate Rohullah. He clearly perceives this dynamic. All the states of subjectivity that he described to me through the narration of his personal vicissitudes have been at times alternatively endorsed and rejected by him, fully or partially.

I will describe a final incident that Rohullah got involved in, in order to corroborate the interpretation that I am giving of Rohullah's inner dynamics. The event took place in Kabul a few weeks prior to my final departure from Afghanistan in June 2013. Interestingly, but I believe not by coincidence, the incident happened when Zair and his family had already moved to the house in Kabul where all the Mangal family lives, including Rohullah. The house is located in an eastern neighborhood of Kabul, where the vast majority of the population is ethnically Pashtun, mostly from the eastern Pashtun provinces of Afghanistan. Unlike the rest of Kabul, which is by and large a "Persian" city, this area of town feels like a homogeneous Pashtun urban environment, where language, outward appearance, and public behavior seem more akin to, say, Jalalabad or Gardez than to Kabul. Rules of comportment follow accordingly. It so happened that the neighbors of Rohullah and his family, also Pashtuns, decided to add two stories to their house, which is separated from Rohullah's house by a wall—as in most houses in urban Afghanistan. The wall was high enough to conceal anything happening in each other's backyards from the view of the neighbors. The two stories that Rohullah's neighbors added to their house, however, rose well above the separation wall. Those inside the new addition could easily see anything happening in Rohullah's courtyard from their windows. Given that, mostly, women are working during the day in the house and courtyard, this meant that the women's privacy (*parda*) was intruded into. It meant also that the honor and name of the whole family of Rohullah was being put in jeopardy. This is by all means an unacceptable situation among Pashtuns, and more generally in Afghanistan at large. Rohullah and Zair urged a couple of times the head of the neighbor family to do something with regard to the problem, such as applying screens to the windows of the two additional stories, or erecting some sort of covered fence on top of the separation wall. The neighbors, however, dismissed their requests, claiming that it was Rohullah's family's responsibility to do something about the issue. One day, things deteriorated quickly. Rohullah recalls:

My older brother Zair was coming back home from work that day, and met one of the sons of our neighbor in front of their house gate. He started talking to the guy, asking him why they were being so difficult and disrespectful about this problem. Then, I don't know exactly what happened, or what they said to each other. I heard Zair screaming outside, and the other guy screaming back at him. You know how Zair is, he does not have much patience when he thinks that he is being disrespected, he gets upset quickly. By the time I got in the street they were hitting each other. I called out Iqbal and my father, and when they arrived there were two or three other people from the other family outside. They had sticks with them, and when my father tried to separate Zair and the other guy, they intervened and hit my father on the head. When I saw my father being hit, I completely lost myself. I got back to my courtyard, picked up an ax, and went outside. I started swinging the ax, and I think I got someone in the arm. Zair and Iqbal had gotten a hold of a stick too, and were fighting with the other guys. Soon after, other people from the neighborhood came to the scene and put themselves between us and the other family. So we stopped fighting. At least one of them ended up at the hospital, I believe. My father later went to the police and denounced what had happened, so that we would be on the safe side. We have yet to reach a solution to the problem of the wall, and to solve the issue of the enmity [dukhmani] between families that has started with this fight.

ANDREA—*How do you feel about the fight?*

ROHULLAH—*Well, that was good ghairat. I saw my father being hit, what else could I do? And also, these people, they are doing something really wrong. . . . I mean, they are not respecting the parda [privacy] of the women of our family. What else could have we done? Sooner or later we would have ended up fighting, one way or another. I think that it is good when I manage to be aggressive in a situation that really requires it. This is good ghairat. I am proud of my ghairat in situations like these. It makes me feel like a real Pashtun.*

Again, the "cultural training" that he went through during the years of his permanence in the village (or rather, the expression of it mediated by a set of meanings that Rohullah chooses now to endorse selectively) still shapes him profoundly, if unconsciously, and is perceived by him as a "positive" outcome of his previous experiences, as well as an integral part of his being.

Furthermore, while listening to Rohullah recounting this incident, I had the impression that the presence of his brother Zair, the very person who played the important part of role model for much of Rohullah's adolescence in the village, had brought vigorously the "village self" of Rohullah back to life in Kabul. More than a matter of which appropriate performance to display in front of one's role model, the incident might be seen as the overlapping, if temporary, of the multiple states of subjectivity (or selves) of Rohullah in a borderline situation where the space is unequivocal (it is certainly Kabul and not the village), but the circumstances are blurred by the presence of a strong symbol of the village life he previously lived (his brother-cum-role-model Zair). Such symbol may have reactivated a certain modus operandi that Rohullah usually reserves for when he is physically in the village. In fact, the previous incident that happened at Ahmed's wedding, which I detailed above, also took place at a time when Zair had already moved back to Kabul.

There is also some even more "positive" use that Rohullah makes, in a conscious way, of the self he had to construe in order to survive socially in the village. When I left Afghanistan, Rohullah was working for a private research company that contracted sociological surveys for the country's ministry of education. Due to his background in Paktia, he was assigned several projects to be conducted in the province. Rohullah said he used his institutional role to awake the "political consciousness [*syasi pohawe*'] of the youth" (in his words).

People know me, remember me, they remember what kind of person I was when I lived in the village. They respect me, they listen to what I say. I still have a group of friends and supporters [andiwalaan] in the village. Now I use my influence [nufuz] for constructive [mufid] purposes now . . . it makes me feel good.

Being "different" by negotiating the private sphere

Rohullah holds a unique position within his family in one respect: he is the only one among his siblings (either male or female) who did not go through an arranged marriage. He "chose" the woman he eventually married. His parents proposed several other girls to him before he met the one he eventually married, but he always refused, and bought himself time to find someone else on his own. He saw her at the wedding party of a friend of his. She was a friend of the bride. Their families were not related. In a cultural environment in which an arranged marriage between relatives is still the norm in the vast majority of the cases, across all social backgrounds, this represented a further point of distinction. They could not talk at the wedding, when they first saw each other, but she was enrolled in a high school located near his house, and he saw her going back and forth from school. He also saw her often going to the madrasa in the same neighborhood, where she attended religious classes. Rohullah was already in university at that time; she was younger than him. Through friends and relatives, he investigated about her habits, her acquaintances, her routines, her reputation among the people she spent time with, and whether she had been already "assigned" to anybody else. Rohullah's younger sister happened to know the girl. Unbeknownst to Asadullah, she and Rohullah's mother went to see the girl's mother (probably unbeknownst to the girl's father as well). After the positive outcome of the meeting, all together with Rohullah's father went to the girl's family house. In the case of an arranged marriage between relatives, usually the individuals involved are familiar to each other, to each family's history, reputation, and social status. Preliminary inquiries and further negotiations are initiated and carried out by the male members of the groom's family, only to be followed through by more extensive surveys also among the female members. In the case of unrelated families, however, a thorough investigation into the past and current social standing of both families is required, which may happen in part covertly, in part openly. Rohullah confessed that he skipped that part of the "procedure." He was not interested in what kind of family the girl was coming from—he was only interested in her, and if she had proven suitable to him, that would have been enough. He let his father, mother, and siblings do the "research" for him. As subsequent conversations demonstrated, it was important for Rohullah to show himself to me as someone who had proven to be more independent-minded and "modern" than his brothers and sisters, and many of his peers as a whole, for that matter ("I am an open-minded person . . . I don't belong in Afghanistan"). He respected the traditional way of doing things (to a certain degree), and did not object to it, although he distanced himself ideologically from it. He could accept the form, but refused to compromise on the content. After receiving consent from both his and the girl's families, he arranged to meet the girl one-on-one, in a restaurant.

He wanted to be sure that the girl would be someone who could live up to his expectations.

> *We met in a restaurant, in the family section, so that we would not be disturbed. My brother Zair and one of her brothers were waiting in the general section of the restaurant. We talked about ourselves, we asked questions to each other. I wanted to know what she liked to do in her free time, what plans she had for the future. I told her the same about me. She told me what she would not like in a husband, I told her what I expected from a wife. It went well, I liked her. She was intelligent, pretty. We had a similar mindset, we had similar ideas about life. I think this is important for everybody . . . to be on the same page with your spouse, I mean. I encouraged her to finish school, and she said that she wanted to have a husband with a university degree. We decided that it was ok to get engaged. I communicated it to my father, who said ok. Then he went to her father and discussed officially the matter. After the engagement party* [when two future spouses are allowed to meet each other in private], *we met multiple times and explained to each other how our families worked, and the rules in life that we would like to have. At that time I was more conservative than I am now, you know . . . I was coming straight from the village . . . I told her about the way I wanted her to go around dressed . . . I wasn't much for women's rights and all that stuff. She told me that she knew about the problems I had had with aggressiveness and fights, and said that she did not like that. She still gives me a hard time when I end up in some trouble with other people.*

Meantime, life in the family's house in Kabul was not being easy for Rohullah. Relations with his parents were difficult. Since he moved back to Kabul, in twelfth grade, he had been living with his parents, two younger unmarried sisters, and his younger brother, Iqbal. The house, which I visited multiple times, was in all fairness quite small for six adult people.

> *Life in my parents' house was too tight for me. We were six . . . too many people. I had my own room, but it was not enough. I had always someone around. I wanted just to be alone, on my own, with nobody telling me what to do all the time. I wanted to have fun, to have my privacy. There was none of that in the house. My father was pushing me to continue university, and possibly do research afterwards, like he did. But, I must say, he was not overly intrusive. He tried to push me by example, more than by words. He showed me the work*

that he was doing, tried to spark my interest. Well, at the moment, the whole thing upset me quite a bit. I was working with a computer design company back then, alongside attending university. I was a lot stressed. Yet my father's pressure left a mark, and I think I owe it to that if I chose to work with the research company that I am with now. I like it where I work and what I do now.

So, Rohullah feels himself to be "different" from the other average Pashtun young men he knows, not only from the village, but also from Kabul. He stresses this "progressive" mindset that he perceives to maintain, which positions him ahead of many others, in his view (see Chiovenda A. 2018b). Although his family (i.e., parents and siblings) is very important for him, as well as certain customs and traditional values particular to Pashtun society (as we have seen), he very proudly affirms his willingness to be his own person, to have an individuality that he protects against excessive encroachments by others, be it his father or the community at large. He wants to think that he is choosing how to live his life, that he is discerning between aspects of traditional Pashtun-ness that he is actively embracing, and others that he is rejecting, unlike many others are capable of doing. The first point of reference that he maintains in this regard is his two older brothers. Nur Mohammed and Zair (the latter until very recently) have continued to live and work in the village, even after Rohullah left for Kabul, and after Zair got his degree in engineering from Kabul University. They embody, for Rohullah, a more "traditional" mindset and way of life. They are better adjusted to the village life. I had never any close relationship with Nur Mohammed, but I had with Zair. Zair certainly thrived in the village in Paktia (he only speaks Pashto and Dari), and, from what I gathered during my conversations with him, he never regretted having chosen to stay. If anything, now he regrets having been forced to move to Kabul, due to threats to his life that the Taliban leveled against him because of his work as a civil engineer. Rohullah always attributed this capacity to better adjust to traditional life that Zair showed mainly to his temperament, which, in his opinion, helped him to fit into a very aggressive, competitive, and often violent environment. Until they both lived in the village, his two brothers occupied the same house with their own families and were subject to the same degree of interdependence and lack of privacy that Rohullah was suffering in Kabul at his parents' house. Yet he considered them to be choosing that type of life, because supposedly better naturally suited for it, and pitted his own recalcitrant attitudes against their acceptance of the communal life they were conducting. He compares unfavorably this "choice to be traditional" that his brothers showed to his own feelings of independence and uniqueness, that he considers a trait of "modernity" on his part.

In fact, the first years of marriage for Rohullah were turbulent. Before the birth of their daughter, as well as for a while after that, Rohullah's wife had problems in adjusting to life with her in-laws. She did not take very easily the fact of being the young bride, subject to the authority of her mother-in-law, the matriarch. In a fashion that is very common in Pashtun families around Afghanistan; the mother of the groom usually exerts a strong power upon the daughter-in-law. Relationships sometimes may get strained, but the daughter-in-law is expected to silently bow to her in-laws' authority and wait for things to get somehow smoother. Apparently, from the accounts that I received in those years from Rohullah and his father, Asadullah, the new bride was rebellious and ended up often quarreling vociferously with her mother-in-law. To that followed a period of so-called depression, during which Rohullah's wife seemed resigned to a life she did not like. Asadullah blamed his daughter-in-law for it, saying that she was "not well in the head," and that she was causing problems for no reason. Over time, things got better, and Rohullah and his father reported improvements and an apparent normalization of the relationships. Rohullah never blamed the disturbances on his wife, however. He always maintained that his wife, like himself, was too "progressive" for such a "traditional" daily life arrangement, that he empathized with her feelings, and that he was covertly working on changing the situation. Rohullah used the example of his wife's reaction to traditional family life to emphasize the point he was at the same time making about himself: I am a modern man, I have chosen a modern woman as a wife, and we are both subject to a lifestyle that we (at least in part) reject. As it happened with all the many Pashtun friends (with the exception of only one) that I made during the years, I have never met or even seen Rohullah's wife (although I have met his mother, due to a principle in Pashtun morals whereby the rules of parda may not apply any longer to an old woman, without any shame or dishonor attached).

Conclusion

After a few years of unsuccessful struggle and suffering in the context of the village, Rohullah worked on himself to "learn" those principles, and behaviors, following which he could ameliorate his social life there. Of course, part of those moral values and ethical rules he already was aware of— he was after all raised in a Pashtun family. Yet, the degree to which he would have to take and implement those values and ethics in order to adjust to life in the village felt probably completely alien to him. His childhood enculturation into peacefulness and mild manners, his consequent abhorrence for violence and abusiveness, had to clash with the reality of the life in the village. Out of pragmatism, and the instinct for (social) survival, he re-enculturated himself, and resocialized himself in the new reality. I interpret the fact that he managed rather successfully to do so as a demonstration of the (unconscious, mainly) plasticity of one's self. As I argued above, this "alternative" self, better suited to the village, that Rohullah shaped over time has become a legitimate and "authentic" aspect of Rohullah as much as the self he brought to the village from Kabul at first. Far from considering this phenomenon as a pathogenic split in Rohullah's original self, or the proof of the pathologic fragmentation of a previously coherent self, I consider this emerged multiplicity of selves as the demonstration of how our psychological organization and configurations may be fruitfully malleable and adaptable. The multiplicity of subjectivities that Rohullah gave rise to actually may be seen as the adaptive response of a well-functioning psychic system to the changed and challenging pragmatic circumstances of life. The integration of the "alternative" self into Rohullah's full being, and the constructive tapping into it that he carries out when required by the contingencies (like when he goes to Gardez to work on social research), represents in my view a positive strategy that renders past life experiences, both painful and pleasurable, available to be used as building material for a self that as a whole will be more adjustable and resilient across different and unpredictable occurrences in life.

At twenty-six years of age (when I left Afghanistan), Rohullah had reached a degree of internal stability (though not devoid of conflicts and hesitations) that allowed him to navigate better than in the past the management of these conflicting aspects of his personality (whether consciously or unconsciously).

3

Umar

The Making, and Un-making, of a Religious Militant

Prologue

Umar is a thirty-five-year old (as of fieldwork) veterinarian from Jalalabad, the capital of Nangarhar province. Like many in Afghanistan who gained a degree in medicine or veterinary science, he has never practiced the profession related to his degree. We first met in summer 2010 through my wife, who, also as a doctoral candidate in anthropology, was conducting research about carpet weaving and embroidery making among Pashtun women. Umar was at the time the executive director of a local NGO in Jalalabad that worked in the districts with both farmers and rural women to enhance agricultural production and female traditional activities, such as the weaving of carpets. The NGO was funded by international private donors. Due to the strict limitations in movement that women are subject to in a Pashtun rural area such as Nangarhar province, I used to accompany my wife (clad in the traditional blue-colored *burqa*, or *chadowry*) to all the meetings and interviews she conducted. It was so that I luckily met Umar and started a fruitful relationship with him that went well beyond the necessities of my wife's research, and which has continued in earnest to this day.

At that time, Umar had under his responsibility approximately seven men and two women. The men would be divided in groups and would go to the rural districts to coordinate activities with the farmers receiving the funds, while the women (in reality, two girls in their late teenage years) would appropriately interact with village women who were to receive material and training in carpet weaving and embroidery. The atmosphere in the headquarters of the NGO was always extremely welcoming and relaxed. The two girl employees worked in an office separate from those where the men worked, and ate their lunch on their own as well. Great attention was paid to the character of the relationship that men and women had to maintain in the workplace. By definition of "traditional" Pashtun social arrangements,

Crafting Masculine Selves. Andrea Chiovenda, Oxford University Press (2020). © Oxford University Press.
DOI: 10.1093/oso/9780190073558.001.0001

men and women unrelated by blood should never mingle together, let alone when away from the direct control of a close male relative of the women. However, nine years after the demise of the Taliban regime, things were already quickly changing in Pashtun Afghan society, at least in dynamic places such as the provincial capital of an important province such as Nangarhar. "Amendments" to the rule of the segregation of sexes in the workplace started to be occasionally tolerated among more open-minded and better-educated families, or, alternatively, among families whose economic situation had become so dire that a pragmatic exception to the norm became the only alternative to complete destitution. The case of the two girls working for Umar fell in both categories. They were both finishing high school (a rare feature among girls in Nangarhar) and lived in families that had lost most of their male breadwinners to the war(s). The relationship between the men and the girls in the headquarters was premised on the tacit acknowledgment that certain rules of Pashtun social etiquette were being broken, and on limiting the interaction between the two sexes to the minimum indispensable for professional purposes. When interacting with each other, men and girls were mutually respectful and detached, and I never detected, during the many visits that I paid to their headquarters, any sort of flirtatious behavior on the part of either side. Against this background, Umar constituted somewhat an exception. The two girls interacted with him on a regular basis, much more than they did with the other men in the NGO. They had to report directly to him, and he had to instruct them on what to do. They frequently visited his personal office and discussed extensively their activities. His attitude toward them was playful and funny, and they responded positively in the same way. I did not perceive any change in such attitude between the times when I visited his office with my wife and those when I visited him alone. He, more than once, explained to us that the reasons why he was taking so many liberties with the girls in his office was that he was playing the part of an older brother with them. There was nothing malicious or flirtatious in the way he conducted himself with the girls, he explained: he only wanted to spur their responsiveness and ambition to improve in their work and, ultimately, life. I must say that nothing in his behavior then, and in what I came to know of him later, ever brought me to doubt his words in this regard.[1]

---

[1] My wife and I often noticed that a way to rationalize, and render culturally legitimate, a relationship that otherwise would stand out as socially inappropriate (like that which Umar was having with the two girls who worked for him) was to create a relation of fictive kinship. The practice of considering the two girls as his younger sisters construed them in fact as *muharram* (i.e., blood relatives,

Umar was clearly a well-educated man, who spoke very good English, and insisted in doing so with me, rejecting my attempts to speak Pashto with him. The first impression of him that I recall is that he seemed the most "western-minded" and open-to-the-"outside" Pashtun man I had met until then. While this was in some regards true, I would later discover that there was much more to it, a more conflictive side of the story.

I kept going back often to the headquarters of the NGO, where I made many other friends, some of whom later became good informants as well for the full duration of my fieldwork. The NGO itself, however, despite the success of its activities (it was one of the few foreign-funded local NGOs that actually managed to create some sort of well-trained workers and self-sustainable business, such as the production of cooking oil and plastic containers for it), lost its funding in mid-2011, and all its personnel had to find themselves a new occupation, including Umar. Part of the problem, I was informed, lay in the fact that the NGO started to be targeted by threats on the part of the "Taliban," who accused the locals working for it of conducting un-Islamic activities, such as spreading Christianity and promoting intermingling of the sexes in the workplace. In late summer 2010, after one of several threatening letters they received from the militant movement (of which I still have a copy), and in order to defuse the tension, the whole crew transferred for a couple of weeks to the Hazara-majority province of Bamyan, to follow more closely the projects they were managing there with potato farmers. My wife and I accompanied them to Bamyan, where we witnessed the tense atmosphere within the group, torn between the necessity to continue working for the NGO and the dangers entailed by it. Sure enough, a couple of weeks after coming back to Jalalabad from Bamyan, the compound where the headquarters of the NGO were located suffered a nighttime attack by the Taliban, which injured nobody, yet shook the personnel to the core. I guess that was the last nail in the NGO's coffin, and funds stopped flowing by May 2011.

I have met with Umar numerous times over the years. During my fieldwork he lived in Jalalabad, and I used to see him on a regular basis, in his office, despite my being forced to adjust to his full, businessman-like daily schedule. When he moved to Kabul, after losing his job at the NGO in Jalalabad, I met him every time I would pass by the capital. We usually had

and hence not susceptible of engaging in any sexual relationship with him), and put them de facto under his responsibility, as to their propriety of behavior and personal safety. Umar represented a guarantor vis-à-vis the girls' family members (with whom he met upon hiring them), and the other co-workers in the NGO.

lunch together during the break at his new job post, where he was subject to an even more hectic daily schedule. Our conversations ran in a less "official" and canonic fashion than the interviews I had with other informants in a "person-centered" way. Thus, in the end, I am not able to say exactly how many hours I have sat with him, one-on-one, talking about his personal life and private issues (many, at any rate). As it happened with Rohullah, he chose to talk to me in English, which I often asked him to integrate with the Pashto rendition of some most interesting passages of what he recounted.

The power of identification, the seduction of power

Umar was born around 1978, in Herat, a major city in western Afghanistan, to a Jalalabad-born Pashtun man, who at the time was an officer in the Afghan Air Force. His mother was a close relative of his father, and was illiterate. When Umar was approximately six years old, his father was transferred to Jalalabad, where he finished his military career a few years later. Umar attended school and university in Jalalabad. The school years, however, were turbulent. From 1989 until 1992 Jalalabad saw furious battles between president Najibullah's government forces and the mujaheddin fighters. Najibullah managed to hold on to Jalalabad long beyond the Soviet troops' retreat in 1989 and Moscow's eventual cut to all financial and military support to his government (in 1992, after an agreement to this effect with the United States—the Soviet Union had just collapsed). Yet, in the end, Najibullah had to relinquish his post, and Jalalabad fell into the hands of the mujaheddin fighters (more precisely, to Gulbuddin Hekmatyar's Hizb-e-Islami party). When the fighting started in Jalalabad, Umar was approximately eleven years old, and he was fourteen when the mujaheddin took control of Jalalabad. He reports the memories he has of those years in Jalalabad as traumatic. Until the dissolution of the communist state, in 1992, his father worked for the air force, and was mostly absent from home. He would see him rarely, mainly at night. When the rockets started to rain on Jalalabad, in 1989, civilian life was hit to the greatest. Umar remembers that streets were deserted, most shops were closed, and only military vehicles were visible in town. Supplies quickly disappeared from the few shops that were kept open, and his family experienced complete lack of food several times for two or three days in a row. His father, with the help of friends, built an underground shelter for his family below their house. Sometimes, Umar remembers, they spent entire days inside the shelter while bombs were exploding around the house. Despite the dangers and the fighting, the fact that Jalalabad continued to be firmly in the hands of the government meant that the most basic services kept on running. Among these was education. Umar's father had put him into a public school, although madrasas (religious schools) were also available and well attended. His father was a secular and mild-mannered man who was absolutely opposed to religious education. Attendance in school was erratic because of the vicissitudes of the war in the area. The state, however, guaranteed the passage to the next class each year to those who attended, regardless of how much of the curriculum had been covered in class. After the takeover of Jalalabad by the mujaheddin, in 1992, public schools were kept running, but their curriculum was shaped after that of the religious schools. After some time into the mujaheddin's rule, the public school system failed altogether, and Umar stopped attending, although he is not sure when exactly. Students who were enrolled in public schools were required to start attending madrasas, which he did.

At this point, the first element of an important personal transformation that will mark Umar's whole life materialized. He started to do very well in the religious subjects that the madrasa he was attending focused on. He was good at the Arabic language, at memorizing the Qur'an, and in the discussions that took place in and outside the classroom about the religious texts they had to read. Umar says:

I was doing well in school, at the madrasa. I was smarter than the other boys, I excelled at studying. I was starting to take pleasure in studying those religious subjects ... it made me feel accomplished.

ANDREA—*Did you feel like you had found a faith that you previously did not perceive? Did you "believe" in what you were studying?*

UMAR—*Well, I don't know, honestly. I don't remember whether I was really, sincerely taken by that stuff back then. I think I did not ask myself this question. It was a strange feeling. If I did good, I would receive a lot of praise and respect. It was like doing something that made me important in the community, among the people I lived with, my friends, my teachers, also the elders. Being a good Muslim was something that was taken for granted. . . . I mean, it was taken for granted that being a good Muslim was something that everybody should want to accomplish. I was showing myself to be a good Muslim, at my young age, and they treated me like an important person. I remember I was proud of it, and happy too.*

ANDREA—*What do you mean by "important"? Did they make you feel "special"?*

UMAr—*Yeah, special, I guess so . . . for example, back then there were a lot of mullahs that were coming to Jalalabad from Pakistan and some Arab countries. They were treated with a lot of respect because we thought they had much more religious knowledge and education than we had. They would stay in Jalalabad, and teach classes in several madrasas. My teachers would introduce me to them, would tell them that I should speak with them directly, because I was a good student. Another example: the leading mullah of the madrasa that I attended one day summoned my father and told him that, in his opinion, I should be sent to Pakistan or India to continue religious education. I was smart, he said, and I needed a higher level of education. Those madrasas were the Deoband- and Salafi-type madrasas* [fundamentalist religious schools]. *My father knew that, and forbade me to go. He wanted me to go to university as soon as I would finish with school education. Also, I would go often to mosque, and sometimes I would be allowed to lead prayer there. This was a great honor for a young guy like me, and my relatives, my neighbors, everybody was really proud of me. It all made me feel good.*

ANDREA—*So, it was more about the benefits that you gained from your good performances . . .*

UMAR—*Bah, I don't know, it is difficult to say. You know, we lived in a very one-sided world. When everything around you is in one way, it is difficult just to step back and do something different. We were completely immersed in that world, we did not see anything else. . . . I was very young. . . .*

Thus, Umar got caught very early in the religious radicalization that the civil war brought about in Afghanistan after the collapse of the communist regime. The mujaheddin, as we heard in chapter 2 from my friend Rohullah as well, were not much different from what came later in the guise of the Taliban regime. What is interesting in the case of Umar, however, is the path that he chose to become a "respectable" Pashtun. Unlike Rohullah, who lived in a much harsher and traditionalist environment, and who, perhaps for this reason, perceived that the way to rise to social distinction was that of playing actively the culturally relevant role of the tough bully, to Umar, who lived in the much more "sophisticated" Jalalabad context, a different avenue of expression and realization became available—that of religious knowledge and performative excellence. In a dynamic and "intellectual" environment such as Jalalabad (which hosts the second oldest university in Afghanistan), the role of the tough and aggressive man was clearly not the only path to affirm one's masculinity and manly worth. As past ethnographic research has indicated (Edwards 2002, Ahmed 1991, Lindholm 1986, 1993), religious charisma has been historically a source of prestige and status among Pashtuns on both sides of the Durand Line (that is, in what is now Afghanistan and Pakistan). The most striking examples of such phenomena may be found during very specific historical moments of upheaval and revolt against a constituted authority that was perceived as inimical in religious terms (like the British Raj, or even a different, but politically dominant, Pashtun group, see Ahmed 1991). In a context of relative sociopolitical "equilibrium," on the other hand, it has been likewise noted that the social standing of the religious leaders was traditionally less relevant, with "secular" local leaders holding more authority than the religious ones.[2] In the context of post-1978 communist Afghanistan, the rhetoric that had catapulted, in particular historical moments, the religious leaders to the forefront of the social and political unrest (i.e., the narrative of the *jihad*, or holy war) was quickly and successfully resuscitated by internal actors through the instigation and support of external forces, both in ideological and material forms (the United States, Saudi Arabia, the Gulf States, and Pakistan). Through this lens we can better interpret the reference that Umar makes to the Pakistani and Arab Islamic scholars who flocked to Nangarhar during his childhood and adolescence, and the revived social status and authoritativeness that they, and their local acolytes, enjoyed. The same trend, incidentally, has not been inverted yet, and, if anything, it has deepened, thanks to the continuing profound influence of the past Taliban regime in the lives of today's Pashtun Afghans, the protracted state of foreign occupation that the country is still subject to, and the consequent, persistent interference of external actors into Afghan political dynamics (mainly Pakistan and wealthy Arab states).

---

[2] Akbar Ahmed (1980) and Charles Lindholm (1996) have in fact taken issue with the portrait that Fredrik Barth (1959) paints of the role and influence of the category (or "caste," as he puts it) of saintly men in Pashtun society. Ahmed and Lindholm pointed out that the degree of political and social clout that individuals with a religious "pedigree" held, showed to be in fact very limited in their experience, and that Barth might have misunderstood this specific aspect of sociopolitical relations within his context of research. As the account that Umar gave me shows, however, this state of facts might have radically changed due to the wars, and the massive influx of ideologies, and religiously minded charismatic individuals who flocked to Afghanistan during and after the struggle against the Soviet occupation. Corroborating this hypothesis, the middle-aged informants I spoke to, who grew up before the Soviet invasion, did reminisce about a prewar period in which the "untrustworthy mullahs" were kept "in their place" by the more powerful community elders, on the premises of customary practices (*rawaj*).

Umar's father, whom he used to see rarely, was apparently unaware of the
steady religious radicalization that his son was experiencing. In this regard,
Umar says:

*My father did not fully understand what was happening in me. We did not
talk very much. He did not know what we used to talk about with my friends
[i.e., religious issues]. He saw I was very pious, very observant of religious
practices, and he knew about the good reputation that these things gave
me within the community. He was not very concerned about it, I guess he
thought it was just like . . . doing things . . . you know . . . things people do to
be considered good Muslims. You know, this is also a problem that we have
here . . . people judge you from what you do in public. If you go five times a day
to the mosque, or they see you pray five times a day . . . if you fast, if you have
a beard, if you wear a skull cap, in other words if your outward behavior is in
line with the rules, they will think you are a good Muslim, they will talk well
about you, they will respect you more . . . but it's only the outside . . . people
don't worry about the inside. If you don't keep this outward appropriate behavior, they will talk badly about you, even if you are a good Muslim [inside].*

The last remark ("even if you are a good Muslim") already anticipates one
major realization, and crucial change in personal beliefs, that Umar experienced after the fall of the Taliban, and which marks him to this day. We will
come back to this point.

Approximately one year after the fall of Kabul to the Taliban in September
1996, Umar started attending university in Jalalabad, in the veterinarian
medicine school. The Taliban kept the university working, and he became an
active part of the movement.

*In university, I became more active as a Talib [a member of the Taliban].
They were controlling all activities, and I joined them, I accepted their way
of interpreting Islam. I know now that it was a very narrow way [tang]
of interpreting religion, but that's what we were given at that time, and
I embraced it [ma da manalo]. I remember it made me feel fulfilled, it
made feel a good person, at peace. I was continuing to do well at the Islamic
subjects. They told me that I should become a tablighi [proselytizer, mis-
sionary], and for that reason they told me I should learn English. I liked the
idea, so I started studying English. It wasn't difficult . . . we could get books
from Pakistan very easily, and there were many Pakistanis among us who*

knew good English. That's how I started with English. . . . I guess it was a for-
tunate coincidence [khosh bakhtana]. *The plan was that I would learn good*
English, and then go outside Afghanistan to preach the ideas of the Taliban.
In the meantime I had also become a preacher inside the university. They
liked that I was good at arguing things, so they made me talk to people to
convince them of the good ideas of the Taliban.

ANDREA—*Were you ok with these activities?*

UMAR—*Yes, I guess yes, I was very much into it. I felt that everything was in its*
place for me.

During our conversations, Umar showed me old photographs of himself
as a Talib, with a black turban, a long beard, and a Kalashnikov rifle on his
shoulder, intent in his militant activities. It was quite difficult to believe that
the person that I was looking at in those pictures was the same one that I had
in front of me, so seemingly "progressive" and open-minded.

Umar's younger brother made a similar choice in life, but of a different kind.
He also joined the Taliban movement, but as a fighter, rather than as an "intellectual." The avenue for expressing his sense of masculine valor and worth
was alternative to Umar's, and still culturally relevant within the Pashtun
Afghan context. After the fall of Kabul, his brother continued the struggle
against the Northern Alliance, the anti-Taliban coalition led by Ahmed Shah
Massoud that held the northeastern corner of the country. His brother was
two years younger than Umar, and this might have influenced his decision.
He set off to the north, where the Taliban were battling Massoud's forces. One
day, after months that his brother had been away, some Taliban comrades
brought him back in a state of shock, after having been involved in an explosion, which did not injure him physically, but shook him psychologically.
He was kept in the family house, and was regularly subject to "psychotic"
incidents, in which he became extremely violent, destroyed furniture and
things in the house, tore apart his clothes, and had to be restrained by force.
Sedation solved his predicament only in part. Over time, his brother slowly
recovered, and returned to be able to participate in social life. His brother's
experience, and his trajectory after his recovery, affected Umar profoundly.

At that moment, however, his brother's trauma did not lead Umar to a
rethinking of his personal path. He says:

I continued to be extremely involved with the movement [the Taliban].
I endorsed their way of seeing religion and politics. I was brought to hate

foreigners, who for me became only kafirs [infidels], *without difference among them. If I had gotten a hold of a foreigner back then, I think I would have killed him on the spot. Foreigners for me meant practically Christians and Jews, who I was told had fought against Muslims for a long time, and, especially in Afghanistan, had worked to control the country, and prevent its people from deciding what to do with their lives.*

However, small cracks began to open into this seemingly unshakeable faith that he was harboring.

One time, during my university years, I decided to trim my beard shorter than I usually kept it. With this shorter beard I went to campus, as I did every day. There, I met a friend of mine, who was a mullah, and also worked for the intelligence in the Taliban movement. He was extremely surprised and aggravated by the fact that I had trimmed my beard. We started a heated argument about whether a good Muslim should be allowed to keep a short beard, or should let his beard grow at least one fist-long. I said that I had read the scriptures, and there was nothing in them which prescribed exactly the length of a beard. He did not want to reason with me at all, started calling me a kafir [infidel], *and said he would get me arrested. I got worried, and contacted a friend of mine, who was a hafez* [someone who has memorized the whole Qur'an], *and was very much respected for that. Through his intercession, I avoided being arrested, but since then I lost any faith in mullahs and people who preach for work.*

Umar, nonetheless, continued to be a part of the Taliban movement, and to prepare for his proselytizing task. In November 2001, at the beginning of the Northern Alliance's offensive, supported by the US military apparatus, he was in Torkham, a border village in Nangarhar between Pakistan and Afghanistan. When the bombs started raining on the Tora Bora area, near Torkham, and on the village itself, he found himself stranded, confused, and not sure about what he had to do. Despite my insistence, Umar never wanted to give me the details of what happened to him after the beginning of the assault on Tora Bora. He only said that after a few days he headed back to Jalalabad, some 75 kilometers away, and went home. My impression has always been that it all took place rather quickly. Notwithstanding this apparent "retreat" on his part, he admitted:

I was in Torkham with my Taliban comrades. We were all there to fight for the cause, to die for it. I was ready to blow myself up for what we had been taught to believe in. To me, at that point, committing suicide for the jihad was the only sense in life. I was still in this state of mind when I went back to Jalalabad. Yet I did not know what to do, the whole thing collapsed, I found myself alone and did not know what to do. I just went home. After a few days I went to the Spin Ghar hotel [the only state-owned hotel in Jalalabad, which was also used sometimes as a rally-point by people] *to see if I could find someone I knew there, and I found it was already full with foreigners. They were mainly journalists. I befriended one of them, and started working for him as a translator.*

ANDREA—*But you told me that you hated foreigners, and if you had met one you would have killed him. How could you decide to help a foreigner now?*

UMAR—*Yes, I know, but I was left with nothing, I had nothing left in me. . . . I had to start everything from scratch . . . the whole system had collapsed, and I had to do something for myself. . . . I did it for practical reasons.*

ANDREA—*Can you tell me about your work with this journalist?*

UMAR—*Well, he was from Britain, I think. He wanted to go to Tora Bora, to see where the Taliban were hiding, after the end of the bombardment. I knew the area, and I accompanied him. We went all around that place, we talked to people in the villages, but there were no Taliban there anymore, they had already left. They went all to Pakistan. I knew that . . . some of my friends did so.*

ANDREA—*What were your feelings while working for this foreigner?*

UMAR—*I had never met a* [Western] *foreigner before. I had only heard what our Taliban leaders told us about foreigners. I believed in what they said, I thought I did not like them . . . but I was also curious. . . . I wanted to know what one of those kafirs was like. All in all, he was a nice guy, he treated me well, he paid me for the time I spent with him going around the districts. It was only for a couple of weeks, but it was interesting. It was good money too. At that time, nobody had any idea of how to keep going on, how to find money for our families. It was fortunate for me to find him in that moment. Then, through him, at the Spin Ghar hotel, I was introduced to a foreign NGO that had just started working in Nangarhar for humanitarian purposes* [insani ʿaraf]. *The journalist was happy with the work I did for him, so he recommended me with this NGO, and they hired me as a translator. They needed people who spoke Pashto . . . nobody could speak Pashto among the foreigners.*

ANDREA—*So, you changed your attitude toward foreigners?*

UMAR—*Well, at that time I felt really very depressed. I still thought that the only meaning in life was fighting the jihad, and sacrificing* [qurban kawal] *yourself for it . . . and that thing was gone, it was not possible anymore. . . . I was confused, I did not see any other reason for living . . . but I had to keep going on, I had my family, I had to help them. . . . I started working in the office of this NGO, and there were many women that worked there with me. . . . I had never interacted with a woman before* [other than his close blood relatives, that is]. *The relationship I had with these women was very respectful, they told me about themselves, and I told them about myself . . . there was this one girl, I think from Sweden, or the Netherlands, I had a closer relationship with her than with the others. I told her that I felt depressed, that I felt that there was no reason left for me to live, and she talked a lot with me about it . . . she told me that I had a lot of skills, that I had still a lot of opportunities in life, that I should not throw my life away like that . . . she convinced me that my life was not over, that I could live for something . . . it was very good for me to meet this girl . . . she changed the way I looked at things. I kept working for that NGO for months, and then for UNAMA* [a United Nations organization], *also in Jalalabad. Over time I realized certain things about my past, and the environment I was raised in. I thought about all the things that they* [the Taliban leadership] *had told us in school and university . . . that kind of Islam was not really religion, it was politics . . . it was political Islam. They distorted* [badal kawalo] *Islam to fit their political goals, and took advantage* [faida akhestale wa] *of the discontent of the people to turn their political goals into religious ones. The people thought they were doing a religious jihad, but it was only in the political interest of these leaders. Now I understand what they were doing.*

A closer look

The personal trajectory that Umar described so far lends itself to more than one interpretation. The oldest child of a military father, albeit apparently devoid of an authoritarian personality, he was confronted very soon in his adolescence with radical religious propaganda. He seems to have embraced with conviction such rhetoric, and to have construed around it a coherent self-image (though not completely in line with the personal representational world that his father would have probably liked him to construe for himself, namely, that of a respectful, but almost "secular," Muslim man; see details in the next chapter about the concepts of self-image and representational world). The mujaheddin took Jalalabad when Umar was still about fourteen years old. To Umar, who was raised in an environment filled with uncertainty and violence, and who was deprived of a strong and present father figure, the mujaheddin years and the subsequent Taliban regime may have brought a modicum of consistency and coherence that he felt necessary at that point in his life. Not having had any other life experience, other than in Jalalabad within his familial context, his horizon toward the "outer" world must have been necessarily limited. Under these circumstances, it may come as no surprise that he promptly endorsed what the mujaheddin, first, and the Taliban, later, provided him in terms of ideological propaganda. Additionally, Umar discovered very soon that his accomplishments in school at religious subjects brought social status and distinction. In other words, he acquired a solid and increasing social capital from actively embracing the new, and successful, apparatus. As we have seen in the case of Rohullah, power, influence, and respect go together hand in hand, and in a competitive and unstable social context in which these features bring a better and safer life (both at the personal and the familial level), it would be comprehensible that, once he found a way to attain these advantages for himself, Umar chose, if unconsciously, to cling to them and their source (i.e., his accomplishments in religious training). Yet the process of personal identification with a religio-political movement and its "just cause" (i.e., the construction of a new self-image, imbued accordingly with novel, subjective meaning), on the one hand, and the narcissistic seduction that the social capital deriving from his choice brought along, on the other, are not necessarily exclusive and contrasting phenomena.[3] Umar, due to his young age and the turbulence of the historical period in which he was being brought up, might have experienced both phenomena at the same time. The combination of the two "explanations" may make sense of the sudden "switching of sides" that he carried out as soon as it became clear that the Taliban cause was lost. He went from wanting to kill foreigners for the sake of jihad, to helping one (the British journalist), and taking his money

---

[3] I intend the concept of narcissism much in the way Heinz Kohut proposed, namely, as an omnipresent, functional feature of human personality structure and dynamic. In this reading, narcissism is something "normal," so to speak, which all humans are subject to in different degrees, and which may present ramifications for psychic disturbance only in certain cases. Such a kind of interpretation of narcissism is far removed from the classical one that Freud proposed, which interprets narcissism as a (mostly) negative distortion of the "normal" psychic functioning.

in return. At a moment when many of his Taliban friends in Torkham took to the mountains in Tora Bora and fought against their enemies, he faltered, and decided not to participate. His friends must have been as aware as he was that the situation was desperate, and probably outright lost. Were they more ideologically committed than he was? Was he more opportunistic than they were?

Such morally laden terms (commitment, opportunism) in my opinion obfuscate, more than illuminate, the phenomena that we are trying to explain with regard to Umar's life choices, especially from a psychodynamic standpoint. The main *fil rouge* that I see in Umar's story is evidenced precisely by his choices to embark in a "career" of thinking instead of fighting among the Taliban, as well as to renounce his cause when it came time to pick up a weapon and shoot. As I mentioned above, in the historical contingencies of Afghanistan in the late 1980s and early 1990s, as it had happened in past circumstances, an avenue open for the Pashtun man who wanted to affirm his masculine worth in a culturally accepted manner was that of becoming a religious leader, which Umar chose for himself. The "alternative" self-image, and set of subjective meanings, that he created for himself (i.e., the Talib preacher), and which he stuck with in order to serve his deep-seated need of belonging to a superordinate cause, as well as, perhaps, to satisfy a narcissistic pull to emerge from the crowd, worked well for him until it became too costly to maintain it.[4] Yet, just like in Rohullah's case, this does not mean that such an alternative self, construed to better adjust to the circumstances of a specific stage in life, should be considered as a "false" self (in a Winnicottian sense. Winnicott 1960), and dismissed as preposterous. In the Umar who wears a black turban, sports a long beard, and goes around preaching radical Islam to his friends, there is a degree of the "legitimate" Umar, which he has retained to this day. Stating that during those years he was "faking" a true religious conviction is to reduce his inner trajectory to *post-factum* moralistic judgments, which do not help a full understanding of his personal path. He has indeed remained to this day a pious, but open-minded, Muslim. The interesting phenomenon that seems to be emerging here is the way in which past and current life experiences, and consequent meaning-creation processes, are managed and elaborated (see Bruner 1990). Through the construction and truthful inhabitation of multiple versions of the self, his "shifting selves," they

---

[4] And in this we can, I believe, clearly perceive the strength of unconscious "pulls" toward self-interest and survival, almost in an evolutionary perspective (see Slavin and Kriegman 1992): a "life-instinct," it would seem, in lieu of the classic Freudian/Kleinian death-instinct.

are alternatively either "stored away" in the unconscious, or pulled back from it and actualized, in relation to the pragmatic necessities of self-interest, (social) survival, and cultural contingencies. We have in fact seen in the previous chapter how such positive effort at alternatively dissociating and linking these self-states, if performed successfully, will contribute to the creation and maintenance of the illusion of wholeness (Ewing 1990) that is crucial for the psychic equilibrium and functionality of the individual. Wilfred Bion (1962) has described powerfully the psychic dynamics that undergird the unconscious processing of "raw" emotional and cognitive material. Following Bion's insight, contemporary psychoanalysts have elaborated on the pragmatic use of such unconsciously "stored-away" material (see especially Ferro 2005, 2006a, 2011, Ferro and Civitarese 2013): the cacophony of the emotional, unsymbolized "ingredients" that wait to be "cooked in the kitchen of one's unconscious mind" (Ferro 2006b) is eliminated precisely by their symbolization, their "cooking," and "alphabetization" (Ferro 2006b. In Bion's terms, this means transforming beta-elements, or emotional "noise," into alpha-elements, or accessible and "thinkable" material—hence *alphabetization*, yet intended not simply as the process of creating verbal thoughts). This process of bringing to awareness and "thinkability" such previously "unformulated experience" (Stern D. B. 2010) makes these states of subjectivity accessible by the individual, in a conscious way, and possibly contrasted with others that potentially clash with them. When the dissociation of these shifting selves is not absolute and defensive, that is, when it is *not* enacted in order to protect the individual's psyche from an unbearable trauma that would ensue from the mutual recognition of these contrasting subjectivities, then the dialogue between them becomes fruitful and enhances the sense of wholeness of the self, contributing to psychological growth (see Stern D. B. 2010, Bromberg 1996, 2011, Slavin and Kriegman 1992, Stolorow, Brandschaft, and Atwood 1987: 29–46, Stolorow and Atwood 1992: 29–40). For many, these psychic processes will happen through the workings of one's own mind, and with the help of daily interpersonal relations, that create an intersubjective "field" within which certain emotional content can finally become formulated and "cooked." I think that an individual like Umar, just like the other characters that we encounter in this book, was able to a significant degree to operate in this way onto his own psychic processes, and to obtain a healthy degree of conflict as a byproduct of an "open," not foreclosing, dissociation. Not that such processes will happen necessarily in a seamless way and without psychological pain. We have seen what degree of disorientation and confusion followed

the choice of going back from Torkham to Jalalabad instead of fighting alongside his friends in Tora Bora, which were slowly overcome only through the
help of insightful external "counseling" (the Western woman at the NGO).

The craft of keeping truthful to oneself

Umar is now the financial manager of a prestigious international economic
organization based in Kabul. After working for one and half years for the
Afghan Ministry of Agriculture, he was hired in late 2012 by this organization, thanks to his long experience in managing funds and his outstanding
record. He married in 2006, and has two daughters. In spite of his "modern"
outlook and demeanor, Umar has a surprisingly traditional marriage history.
He married one of his cousins from Jalalabad, an illiterate woman who was
forced upon him by his mother. His mother is also illiterate, and apparently,
from Umar's account of her, a strong-willed woman, with very "traditional"
views about family life. Umar did not want to marry his future wife, and tried
all he could to convince his mother to accept that he choose someone else for
himself. Yet there was nothing he could do about it, and his father just went
along with Umar's mother's wishes. Umar says:

> I tried for a long time to suggest to them [his mother and father] *another so-*
> *lution. Me and my brother, we did find a few other girls that might have*
> *been better for me. They were better educated, they had gone to school, and*
> *I wanted someone who was at least a little bit educated. I did not want to*
> *have an illiterate wife. But my mother imposed herself. She did not accept*
> *any of the girls that I proposed, and chose one of my cousins for me. I could*
> *not change her mind.*
> ANDREA—*Why did you just let her choose for you, in the end?*
> UMAR—*Well, you know, she's my mother, that's my family, I did not want to*
> *create a bad relationship between us. I could not go against her wishes.*
> *I would have created a fight between us, I would have been left on my own,*
> *they would have refused to help me. That's how it goes here in Afghanistan,*
> *you are expected to respect your family's choices in these cases.*
> ANDREA—*But why do you think your mother wanted you to marry just this*
> *one girl, someone who had no education, and not another one? You are a*
> *veterinarian, many other girls would have married you.*

UMAR—*I know why . . . my mother is a powerful person, she is strong . . . she wanted someone whom she would have control over, after the wedding . . . she wanted someone who would do whatever she told her to do, without making problems. That's how it goes here . . . your wife becomes the servant* [gholam] *of your mother, she takes orders from her . . . she does not have to make problems. My mother knew that if I had married someone with an education, she would not have accepted to live under her control.*

ANDREA—*So, that is how it worked between your wife and your mother?*

UMAR—*Yes, in the beginning. Then we had kids, and the situation became a little better for my wife . . . but my mother is still very strong with her. Thankfully, now we live in Kabul. . . .*

ANDREA—*How is your relationship with your wife?*

UMAR—*It's ok, it goes well, we don't have any problems. She stays home, she looks after the kids, I stay most of the time outside the house. . . .*

ANDREA—*Are you satisfied with your wife as a partner?*

UMAR—*Well, I can't talk with her about much . . . she is uneducated, I can't talk with her about things that I would like to discuss with a wife. I would like to have a more interesting person with me. But that's how it goes, what can you do. . . .*

ANDREA—*So, you are not happy?*

UMAR—*No, I am happy, I am. . . . You see, after some time, you just accept what you have, and try to make the best of it. This is how I dealt with the problem. If you continue thinking about how better it could have gone for you, and all the better opportunities you could have had, you live a horrible life. In the beginning, I was unhappy, because I was always thinking about what I really wanted, what I could have had* [i.e., the wife he could have had]. *I was unhappy, I was not living a good life. But then, after a while, I said to myself: hey, you can't change this, this is what you have, and you will have to stick with it. You better adjust. I knew I could not change this situation, it was useless to complain. So I just resigned myself to it, and started looking at the good things that I had, like the kids, a good job, the chance of going abroad every once in a while* [because of his job]. *I stopped looking back, and only looked ahead, trying to live well. It got better, now I am happy.*

ANDREA—*Well, you could have divorced your wife. . . .*

UMAR—*No, no, are you joking? No, I could not have divorced my wife! No . . . that would have been terrible for the honor and reputation of my family . . . no, there is no divorce: once you are married, you stay married* [divorce is allowed by law, in Afghanistan, for both for men and women, but socially frowned upon].

ANDREA—*Do you go out with your wife? Do you have any outside activities with her?*

UMAR—*No, I don't. I would like to, but she is uneducated, she is not used to be around other people* [outside the circle of close relatives] . . . *she would embarrass me, she would do something wrong* . . . [i.e., she would behave inappropriately with strangers, and compromise the family's name and honor].

Umar ended up doing what most men, and virtually all women, do in a Pashtun social environment still today: he delegated the decision about his marriage to his parents. This phenomenon, of course, is not peculiar only to Pashtun society. Cross-cultural psychoanalysts and psychoanalytically inclined anthropologists have described and analyzed cases of this sort from other sociocultural backgrounds (Frie 2008b, Roland 1988, 2011, Ewing 1990, Stern D. 2001 [1985]). Individuals live and rationalize such state of facts in different ways, idiosyncratic to each of them. The fact that individuals let this happen to them, all the more because the event is often accompanied by a varying degree of internal suffering, has been interpreted alternatively as the symptom of a lack of individuality and autonomous will due to a "sociocentric" enculturation (Markus and Kitayama 1991), as a proof of an "unbounded" self, in opposition to the Western supposedly "bounded" self (Geertz 1984: 126), or even, from an ego-psychological point of view, as a pathogenic merging of selves between the individual and his socially significant others (Ewing 1991). All these interpretations, on the one hand, imply that the self may be so strongly and profoundly shaped by the cultural environment to the point of having *structurally* diverse configurations in relation to different cultural environments, and, on the other, imply in a veiled way also that the "individuated," bounded Western self will provide the individual with healthier psychological dynamics.

I believe that the inner dynamics that my informants have displayed through their conversations with me, including Umar, tell a slightly different story. It is true that the sociocultural context in which Umar, for example, has been raised and lived, has impacted him so powerfully to lead him to put the interests of a socially crucial group for his life (the family) ahead of what he perceived to be his own personal interests (i.e., to have a wife who would match better his personality). However, the wishes of his family members have acquired a pivotal importance also for *his own* set of meanings—it is crucial for him to maintain the love and consideration of his mother and father, and to respect their decisions also when they regard his own personal matters. The positive linkage to his parents has become an important aspect of in his life, and has become a legitimate, "authentic" aspect of his subjectivity; it has become, in other words, part of his broader personal representational world (see Sandler and Rosenblatt 1962, on the overlap between personal and cultural representational worlds, as well as next chapter for a full discussion on the issue, and for my understanding of "authenticity"). This is a choice that he seems to make in an unconscious way. Its roots, which lie in the meaning-creating power of cultural and social constraints and patterns, seem to remain out of awareness. What comes to the surface are the emotional and psychological consequences of the conflict: frustration, anguish, unfulfillable fantasies. It is also true, however, that Umar never loses track of his very private, deep-seated wishes and preferences. He maintains a contact with his emotions and feelings, with a "private self" (Modell 1993) that would lead him toward a different direction had he not developed, through enculturation, a parallel set of personal meanings (a different self-image, another version of the self), to satisfy, which he gives in to the wishes of his mother and father. Umar "feels" his private self, his individuality, and nevertheless chooses to pursue a different path because it appears to him better suited to the pragmatic social and cultural circumstances he has to live in. As a young man living in a traditional Pashtun environment, with all that it entails, it is more advantageous for him to assuage the wishes of his parents than rebel and turn them into enemies. If the latter scenario had ensued, his personal and social life would have been destroyed. Whether consciously or unconsciously, Umar is aware of this delicate balance, and acts accordingly to his primary interest. In the specific case of Umar, I believe that his capability at introspection was good enough to have rendered him at least partly conscious of these processes. So much so that, after a period of disorientation and rejection of the situation ("I was unhappy, I was not living a good life"), he shows to be strong enough to put himself back together, rationalize what had just happened, and work constructively to extract the better possible life from the predicament he found himself in ("I stopped looking back, and only looked ahead, trying to live well. It got better, now I am happy"). His adjustment is remarkable.

There is nothing intrinsically pathogenic in how he conducted his actions to this point, and there is nothing that should make us believe that the impossibility of realizing what he really wanted for himself from a marriage is the symptom of a lack of an "individuated-enough" self.[5] (We will observe in more detail an analogous situation with regard to another of my informants, Rahmat.) And from all of the above, it certainly does not obtain that the possibility to actualize at any given time the imperatives of a "private self" is a necessary condition in order to attain a healthy psychological life. As we have seen also with Rohullah previously, enculturation prompts the creation of specific sets of meanings for each individual, which become legitimately and positively "his own," and that may require a culturally appropriate pragmatic fulfillment. As long as the parallel, complementary sets of subjective meanings created by the individual in response to both private and environmental cues (i.e., one's self-images, or selves), manage to remain, if only partially and unconsciously, in contact, and in a condition of mutual exchange, I argue that the individual will enjoy psychological growth and psychic functionality (on this, and on the positive process of "healthy" dissociation that underlies the "exchange" I am referring to, see Bromberg 2011). It is in this sense, I argue, that cultural and social training impacts profoundly the self of each individual. As the Japanese psychoanalyst Takeo Doi convincingly indicated (Doi 1981, 1986), it is precisely, if paradoxically, in societies that display a very strict set of social norms and rules to be followed, in order for an individual to be validated by one's community, that the "private" self of the individual and its wishes emerge to awareness more strongly, notwithstanding (or because of) the fact that they might become subordinated to socially driven priorities (see also Lindholm 1997, Roland 2011).

Sometimes, during our conversations in his NGO office in Jalalabad, Umar would receive phone calls from women, with whom he engaged in lengthy conversation. I asked him about it. He explained that usually these were Afghan women who worked in other NGOs or government institutions, and with whom he liked to maintain a friendly relationship by phone. They would never meet, he assured me, but they would talk often and at length,

---

[5] I am referring here to the concept of "separation-individuation" introduced by Margaret Mahler in the 1960s (see Mahler 1979).

about various issues—mainly topics that he was unable to talk about with his wife. He then explained:

Talking to these women gives me some satisfaction, and pleasure. You know, there is nothing sexual about it, I do nothing wrong, but talking about interesting stuff with these women gives me some intellectual pleasure [zihni khoshaltob]. I like to have some sort of interesting relationship with a woman, like I cannot do with my wife. You know, things are changing in Afghanistan, talking on the phone with a work colleague of yours, a woman, is possible today, people don't see it as a bad thing necessarily. Well, I can't do it at home, this is true, because my wife gets jealous and gets upset about it . . . but that's because she is uneducated, she does not understand. . . .

ANDREA—*Are the women you talk to married?*

UMAR—*Some are.*

ANDREA—*And their husbands are ok with you talking to their wives on the phone?*

UMAR—*Yes, generally they are, they hear that it is stuff concerning their wives' job, they understand.*

ANDREA—*Do you ever think that you would like to be married with one of these women, instead of your wife?*

UMAR—*Well, yes, I know that I would be better off with someone like them. But, I told you, I am over that thing now: it can't happen, it will not happen, so why get sad about it? I am happy with the relationship that I have with these women as it is now.*

The frustration that Umar feels with regards to his wife and marital life in general is partially eased by maintaining some degree of relationship with other women, if only intellectually. The fact that he now professes to be "happy" with his wife must be still viewed through the lens of the pragmatic choice that he made, in order not to continue living a miserable life, vis-à-vis his inability to change his situation. He realizes the sense of fulfillment that interacting with these women gives him, and yet this pleasurable feeling does not manage anymore to challenge the status quo that he has attained in his personal life; he has cut the competitive link between the two realities. In cultural and social terms, having female "phone-friends" is not at all uncommon in Afghanistan among Pashtuns. Several married men that I know have this kind of relationship, often with more than one woman. They might meet these women (secretly) or not, they might engage in sexual relationships with them (even more secretly) or not. In an urban center like Jalalabad, this is easier to accomplish than in a rural village, for obvious reasons. Also, which is less obvious, the consequences of being caught in a rural village might be more serious than in the urban center (that is, deadly). To know the truth about these kinds of relationships is hard, because men like to brag about it, and acquire social status among their peers through sporting their Don Juan–like stories like trophies. I have been warned about the disputable veracity of some of these stories. On the other hand, however, I have been present several times when friends of mine were speaking on the phone with their female "phone-friends," and have witnessed the flirtatious nature of the conversations. Their wives usually do not know about such schemes (although sometimes the women they engage in conversations with are in turn married themselves, which means that the trick is played mutually by both sexes). Flirtation and outright courtship is often an element in these relationships, whether sex is involved or not (in many cases it is not). Nevertheless, in a social environment like the Pashtun one, where relations between men and women, even if friendly and innocent, are extremely limited (often restricted to one's close relatives), the mere act of talking to a strange woman on the phone may provide an exciting, sensual, even perhaps erotic pleasure to the man involved (and probably to the woman as well). This realization was brought home for me when it became clear that some of the male informants that I had had the habit of calling my wife on the phone, instead of directly me, when they wanted to talk to me (and when my wife was present in Jalalabad, of course). They called my wife to have the chance to exchange a few words with her before talking to me. They tried to know more about her, and generally chat. Had they done that with a Pashtun friend of theirs, that fact alone would have given sufficient grounds for a bloody reaction, against both the wife and the friend. But a foreign woman brings with her no social strings attached: no dangerous gossip, no likely retaliation, no deadly feuds. The ritual of the courtship, which most men do not really experience (being married off by their parents to practically strangers—although technically relatives), is also an aspect of the relationship that men and women painfully know they lack, and which may be pursued through these illicit stratagems.

Maintaining phone relationships with his female work colleagues was therefore for Umar a means to somehow attain the goals I just sketched above, although he was playing on the safe side, because he had a "legitimate" reason to spend time talking to them—common job-related issues. It was for him, as it is for many others, I believe, a way to cope with the inner conflict deriving from being aware of the deficiency in one's life (being married to the "wrong" person), no matter how much pragmatically and rationally Umar, in this case, might have managed to overcome the initial burning dissatisfaction and frustration.

The deep reach of enculturation

Being married to the "wrong" person brings about other related issues. One of the cornerstones in the idioms of masculinity among Pashtuns in Afghanistan, as it happens also in other Mediterranean and Middle Eastern contexts (albeit to a different degree), concerns the control that male family members must exert on the family's women. Since much of the *namus* and *izzat* (honor, name) of the family depends on its women's behavior, their possibility for autonomous action in the public sphere is strongly curtailed (if not altogether obliterated). This phenomenon is taken to the extreme, when, for example, two men who have been, say, best friends since childhood and spend most of their time together still in their adult years, will never see each other's wives, despite the fact that they both visit each other at their respective homes very often. I discussed this matter with Umar extensively in one of our interviews.

ANDREA—*Do your friends come often to see you at your house?*
UMAR—*Yes, pretty often.*
ANDREA—*Do they come with their wives?*
UMAR—*No, only my friends come to visit.*
ANDREA—*So, you never have dinner all together with your friends and their spouses.*
UMAR—*No, it never happens.*
ANDREA—*Not even with your best friend?*
UMAR —*No.*
ANDREA—*Why is it so?*
UMAR—*Well, it is not appropriate* [munasib], *we do not do that . . . people would know about it, they would start talking . . . it would give us a bad name. . . .*

ANDREA—*What would they say?*

UMAR—*Well, they would probably say that I am beghairata* [without honor, without manly attributes], *because I let my wife be seen by others* [i.e., not close relatives].

ANDREA—*OK, I get it. . . . Imagine that you can make sure that nobody will ever know that your best friend has seen your wife at dinner . . . imagine, right? . . . would you allow him to sit with your wife and eat with you two?*

UMAR—*Ummmh . . .* [long pause] *No, I don't think so.*

ANDREA—*Why not? Nobody knows, nobody is going to call you beghairata!*

UMAR—*Well, it's not only that . . . maybe we sit down, all together, and we start talking, and my friend starts flirting with my wife. . . .*

ANDREA—*Your best friend, flirting with your wife?! I don't think it would happen . . . don't you trust your best friend?*

UMAR—*I don't know, anything can happen, maybe they start looking at each other in a strange way. . . . Maybe my wife likes him, and then something starts from there, and I am in a lot of trouble. . . .*

ANDREA—*Do you think that an illicit relationship may start from that?*

UMAR—*Yes, you never know . . . you never know what can happen.*

ANDREA—*Do you think that this has to do with the fact that the marriage was arranged? I mean, you did not know each other, you did not choose each other.*

UMAR—*Maybe, I don't know. . . .*

The fear that his wife could get interested in someone else is at the root of a sentiment, and its related controlling behaviors, that in the West would be commonly called jealousy. Such extremely pervasive and strong feeling of insecurity, precariousness, and imminent loss of a value (his wife) is so terrifying that strict measures are enforced to make sure that the catastrophe will never happen (if she should misbehave, or if he should be forced to divorce her, it would be a great shame for the whole family, as we have seen. Having a family is a social "obligation." In this sense one has to protect the union of wife and husband). A wife will never have to meet anyone other than those whom she cannot marry, as defined by the tenets of the *shari'a* (i.e., the *muharram*, or close blood relatives) lest you invite disaster (in the guise of public shame). And yet, in spite of all the precautions taken, the sense of insecurity persists, and, if anything, seems to worsen, in a vicious circle. Upon discussing this issue with many other friends, my impression is that the fact of having no choice as to whom one will marry, and having often a complete stranger assigned to oneself as a wife, opens the way to the unconscious acknowledgment that there will certainly be somewhere, "out there," someone that your wife will be better matched with. Therefore, the more chances she will have to interact with other people, the more chances she will have to find this better match, which is to be avoided at all costs. Even though Umar was not able to fully articulate this concept, the latter has nonetheless been suggested to me by many others, who behave in the same way with their wives as Umar does, so that I think it may be proper to Umar's psychic functioning as well.

There are several processes in action here, I believe. On the one hand, there is the conscious realization that one has been forced to marry someone that one did not choose, and much probably did not like, and consequently, there exists the longing for someone else better suited for oneself. On the other, there is the unconscious projection of this very same feeling onto the actual spouse, whom one imagines being certainly animated by the same feelings (which may not be necessarily true). The projective mechanism in this case is rendered more powerful by the sociocultural arrangements of Pashtun society: men not only long secretly for someone else to be with (as women as well likely do), but are in fact "unofficially" allowed, without much public scandal, to pursue the search for someone more intimately satisfying than their official wife, with whom to engage in extramarital affairs. Certainly, religious principles forbid such course of action, but, as many have told me, "what can you do, so many do it, we don't make a big deal of it anymore." In addition to this, men can always marry legally multiple women (up to four), so that it often happens that those who can afford it economically, marry "for love" a second wife (or more). So, desire, for men, has an outlet for expression (official or unofficial), and desire is a feeling that men recognize very well, and nurture extensively. Women are not allowed any of these outlets for the expression and fulfillment of desire. They are assigned a man, and will have to be content with him. The actualization and satisfaction in men of a powerful affect such as desire makes them aware of the potential subversive nature of such feeling. The unconscious projection of the same affect onto the subjectivity of their wives (whether with good reasons or not), therefore, has devastating consequences that are partially responsible for the plethora of extreme controlling measures, on the part of the male family members, that characterize the life of a Pashtun woman.

Such practices of strict control over women's behaviors and participation in public life may be interpreted as a "culturally constituted defense mechanism," in the terms defined by Melford Spiro (Spiro 1952, 1965), upon the suggestions first advanced by Irving Hallowell (Hallowell 1955). The painful and uncontrollable sentiments of fear, insecurity, and precariousness that the sociocultural arrangements concerning marriage among Pashtuns give rise to, are tentatively kept in check, or under control, by these practices, that are precisely engineered to defuse and neutralize these feelings. Umar, notwithstanding his "progressive" and open-minded approach to life, is not immune from these dynamics. The position in which Umar stands is not only testimony to the power of enculturation upon the individual. It also seems that Umar acts a reluctant victim of such culturally constituted defense mechanism. He pragmatically accepted, and rationalized to himself the fate that he was assigned (that is, to have a wife whom he does not appreciate). Once put in the condition of many others, however unwittingly, he is prey to the same psychological mechanisms that derive from the objective contingencies, and acts accordingly (i.e., embracing a culturally constituted defense mechanism that results in "obsessively" controlling practices upon his wife's behavior). His masculinity finds expression, in this case, along culturally "traditional" lines.

I cannot know for sure whether Umar has ever engaged in any illicit sexual relationship with any woman in Afghanistan. Based on my knowledge of him, and my experiences with him over the years, I have no reason to doubt what he said to me, that is, that he has never done so. However, when he was still amid the disorientation and the inner turmoil immediately subsequent to his unhappy wedding, he did take advantage of one occasion he had in order to "revolt" against the fate he had been condemned to, and to relieve his marital discomfort. Soon after his wedding, he took a couple of job-related trips to Europe, in the Netherlands and Germany. He confessed to me that in the Netherlands he did have a sexual relationship with a woman he knew there. How he reconciled this with his religious piousness, and with the knowledge that his action would have not been condoned by the letter of the Qur'an, he has never explained. However, such phenomenon of cognitive dissonance is so common, and goes so much unnoticed (or rather, ignored) among many Pashtun men (the quintessence of a pious Muslim man, in their own opinion), that it has never really surprised me to find the same happening to Umar.

Umar has emerged from all the vicissitudes in his life as a self-described renovated man, vis-à-vis the young adolescent who not only got seduced by the rhetoric of a radical religio-political movement such as the Taliban, but actively took part in it. His piety is now lived by him in a very "private" way, a way that he finds more "sincere" than the way in which most of his peers inhabit religion: less audience-oriented, empty performance, and more deep-felt meaning. He has now found the real meaning of life, he told me in one of our last interviews.

> *I live for my children now, and I am happy when I go back home from work. This is what Allah has created us for, not for sacrificing ourselves for the sake of religion. Religion must be lived in a different, positive way.*

Conclusion

In Umar's inner trajectory from early adolescence to adulthood, I believe we can again see the dynamics of shifting selves in action. He construed for himself an "alternative," yet at the same time complementary, self (the Talib preacher), which was probably born out of the (unconscious) need for a superordinate cause with which to identify and participate in, as well as an (equally unconscious) narcissistic urge to stand out from the crowd of his peers, in order to gain broader validation and acknowledgment from his community leaders and family members. The secular familial context in which he had lived up to that moment, characterized by a father who never pushed religiosity upon his children beyond the standard fulfillment of practices that would sanction an "average" Muslim individual, was therefore recused by the young Umar, who embarked into a self-propelled work at radicalization, which took advantage of the cultural and political influences that were filling the social landscape in Nangarhar at that time (beginning of the mujaheddin's rule, and subsequent Taliban regime). Umar, however, who could have decided to engage in a militantly violent course of action, as his younger brother did, engaged instead in a lifestyle that was better fitting a broader inclination to intellectual speculation and emotional balance (hence the religious preacher), which he felt from the beginning, and maintained until the present. In other words, while the cultural and political environment in which he spent his adolescence "cooperated" with unconscious urges to build a strong self that would provide him with both the transcendent meanings and the social capital that he was striving for, certain crucial aspects of his "antecedent" self-image were retained and incorporated in the "new," alternative/competing/complementary one (e.g., reflexivity, lack of aggressive impulses). Such crucial aspects returned to the forefront of his matured subjectivity after he abandoned the robes of the Talib preacher, at the end of the Taliban regime.

During the painful and long work at reconstituting a representational world that would feel coherent and attuned to a changed reality of personal life and sociocultural environment, Umar revitalized those subjective meanings that he had "stored away" during the previous years spent as a religious militant, while at the same time he also retained a part of the self-image he shaped as a Talib (that which he deemed more positive and helpful for the new reality he was immersed into after the fall of the Taliban). In fact, the religious training he underwent during the years under the mujaheddin and the Taliban did indeed profoundly shape him, and his mindset. The need for piety and transcendental meaning that he sought during his adolescence, and which was fulfilled through radical religious rhetoric, was given room and a new form after the demise of the Taliban. A different religiosity pervades Umar today, one that he finds more attuned to his current being, and his goals in life. Upon my experiences with him, I have no reason to doubt the "sincerity" of Umar's profession of faith. The open-minded and "progressive" Muslim man that I have met since the beginning of our relationship is the coherent product, if apparently unexpected, of the trajectory we have seen develop.

The dissociative process that Umar employed to transform himself from the young adolescent, brought up in a fairly secular household, into a radicalized religious militant, never entailed a foreclosure of the unconscious dialogue between these two subjectivities. The dissociation was at time painful, but never absolute and defensive. Conflict emerged as soon as the Taliban regime fell, and the avenue for exchange that he unconsciously always left open between his two self-states allowed him to enter that potentially fertile ground of inner yet conscious conflict that set the foundations for his post-Taliban new life.

The details that Umar gave me about his marital and sentimental life add some interesting elements to this picture. The story of his marriage, and the way he managed the frustrations deriving from it, speak to the power and reach of enculturation and social constraints. Umar cannot escape the force of customs, and bows to the imposition by his mother to accept as a wife a woman whom he does not want. Thus, he does resign diligently to his fate, though not as a symptom of a "pathological" merger (in ego-psychological terms) of subjectivities with his (bad) self-object (as Heinz Kohut thinks of it—in this case, it would be his mother's wishes), but instead as the product of an unconscious "choice" that undergirds the importance that familial harmony and parental love maintain for Umar, within a sociocultural environment in which these values acquire a pivotal role for the psychological balance of the individual. In this sense, Umar is very much the product of his cultural and social context. However, he remains conscious of his dissatisfaction (and the manifest reasons for it), and of his longing for something different. There are no grounds in this regard to postulate a "merger" of subjectivities. To these elements of a "traditional" self, derived from the work of enculturation, Umar nevertheless is able to oppose elements of a more "progressive" self (as he puts it) that he has struggled to gather after the dramatic and conflicting experiences in his life. His detachment from religious radicalism, his new interpretation of religion as a private endeavor, based on tolerance and mutual understanding (almost in "secular," Western terms), his opening to a more "modern" lifestyle (in his opinion), are the product of the interplay between the external inspirations that he was exposed to while working with foreigners after the demise of the Taliban, and his own personal re-elaboration of the unsettling experiences he went through during his adolescence and early adulthood. The emergence of elements of what he considers a "modern" subjectivity are thus the result of a constructive and healthy psychic process, in my opinion: he "learned from experience," to paraphrase Wilfred Bion. It is constructive and healthy not because "modern" per se, but because apparently functional to his current psychological equilibrium.

4

Baryalay

Between Cultural and Personal Representational Worlds

Prologue

Baryalay Saidy was born around 1985 in a remote village in a rural and (to this very day, in 2019) very volatile district in Nangarhar province. His name (Saidy) shows his belonging to a "saintly" family of *pacha*s, that is, of people who claim to be descendants of the Prophet Mohammed. Located approximately a one and half hour's drive south of Jalalabad, toward the towering Spin Ghar mountain and the border with Pakistan, the area where Baryalay was born and grew up has become over recent years more and more fertile ground for the development of a fierce antigovernment and anti-occupation sentiment, which produced an extensive support base for the Taliban movement and its many local offshoots. State control is virtually absent in the district. Ideological influence from radical religious centers in Pakistan is strong, and material support flows freely across a porous border. This state of affairs heavily influenced my relationship with Baryalay: he is the only informant whom I was able to meet exclusively in the "safe" environment of my residence in Jalalabad and whose friends and relatives I was never able to meet. Baryalay was one of my most recent encounters in Afghanistan, yet he is also the informant with whom I eventually held the most numerous interview sessions (thirty-seven) in Afghanistan, between 2012 and 2013—he was my language teacher.

When I arrived in Afghanistan in August 2012, I looked for someone with whom I could have Pashto conversation classes, so that I could improve my language proficiency. A friend of mine in Kabul, Jalaluddin, a bright young man who had recently received a master's degree in communication in the United States, referred me to Baryalay as a person who would be suited for my necessities. I later discovered that Jalaluddin's wife was the sister of Baryalay's wife. What started as a simple language practice relationship soon turned into much more. Baryalay showed himself immediately as much genuinely interested in

Crafting Masculine Selves. Andrea Chiovenda, Oxford University Press (2020). © Oxford University Press.
DOI: 10.1093/oso/9780190073558.001.0001

speaking with me, and finding out things about me, as I was about him. Our sessions proceeded in a much more engaged and "reciprocal" way than I would have anticipated in the beginning. Topics for discussion were never lacking, and particularly Baryalay's willingness to speak freely never faltered. I informed Baryalay of my goals for fieldwork, and my research interests, and he gladly accepted to become a part of my study, while keeping always as a first priority the improvement of my language capabilities. Our sessions were all held in Pashto, although we interspersed our conversations with some English when needed to better explain a specific term or a grammatical construction. In fact, Baryalay was able to speak a decent English, which he, however, diligently (and "professionally," given his main role as language instructor) refrained from using with me as much as possible. He is a well-educated man, who received a BA in agronomy from Khost University, in an area of southeast Afghanistan adjacent to the border with the Tribal Areas of Pakistan (FATA). When we met, he was working for a Western-funded NGO, which supported agricultural development in Nangarhar province. He was a field officer, that is, a staff member who works most of his time outside, touring the different sites that the NGO manages. He liked his job, which put him in direct contact with farmers and community members, and limited to the minimum the time spent behind a desk in an office. His tasks took him far and wide across the province, from the few remaining pine forests (*Pinus wallichiana*) in all of southeast Afghanistan, on the mountains at the border with Pakistan, to the artificially irrigated fields in the hot valleys of the Kabul and Kunar rivers. I always thought that his demeanor and posture would help him in dealing with farmers and land owners, the recipients of the funds whose best utilization he had to oversee. His slender and tall figure, coupled with a soft-spoken and ever-respectful attitude, as well as a professional competence that I discovered over time, made him a person well suited for dealing "diplomatically" and tactfully (hence successfully) with all those problems and inconveniences that might arise when large amounts of money are handed out to individuals and communities that have very little and need very much.

The development of our sessions was interesting. We would meet in the house that I rented for myself in Jalalabad, and talk for a period of time that varied considerably, from a half hour (very rarely) to roughly two hours (more often). At first, I approached the sessions simply as a conversation practice in Pashto, as it was supposed to be. I paid him a flat rate for each time he came, regardless of how long he would stay. I had to insist and impose myself in this regard, because in the beginning he refused to be paid at all. I had anticipated this because it would be disrespectful to receive money for a service done out of a sentiment of hospitality, and in honor of a guest. I therefore explained to him that, in my culture, it would have been shameful for me to receive a service such as the one we were planning without rewarding the counterpart with anything at all. He remained convinced of my explanation, and agreed to be paid. After a few sessions of general talk, noticing the positive responsiveness that Baryalay displayed, and his good disposition at talking to me, I tried to touch more explicitly on issues that were of interest for my research. My attempt surely did not go unnoticed to Baryalay, who was happy to play along. We spoke more and more often about his vicissitudes in life, past and present, and once the conversation took off around a specific issue, I tried to keep my interference to the minimum. In the end, the Pashto conversation class became little more than an excuse to meet and talk about topics that were relevant both for me and for Baryalay (although we did continue to speak only Pashto, and diligently take notice of all those linguistic features that I was not familiar with. Indeed, the constant practice done with Baryalay proved crucial for my acquired proficiency in Pashto). He talked as much about himself, his social life and his personal history, as he asked about me and social relationships in "the West." His curiosity and interest for the "outside" world was a characteristic that Baryalay himself felt very important for his own development as a human being. To this he often contrasted the "ignorance" of many of his fellow villagers, who, in his words, had never left their small world, and had no cognizance of how to live in a "dignified and civilized" manner. We will see that this attitude reflects an important aspect of Baryalay's self-representation.

Among all the informants who worked with me, my sessions with Baryalay are the ones that resemble more closely a psychoanalytic/psychotherapeutic setting. We always met in the same place, at a prearranged time, and for a length of time we would set prior to each session. The continuity and frequency of our meetings were such that after a while I did not have to ask questions any longer at the beginning of each session. Baryalay would come to our meetings and just start talking about things that were popping up in his mind at the moment. He knew that the main scope of the deal was to have a conversation in Pashto, and he understood that I was genuinely interested in anything concerning his life history, his personal current vicissitudes, and their attendant psychological ramifications. So he just talked, without worrying much about the specific subject. Such situation was best suited for the work on free associative thinking that is central to any psychodynamic analysis. I would of course ask questions every now and then, either when I did not understand some passages of his narration, or when I wanted to have more details about aspects of it. Undoubtedly, between a psychotherapeutic relationship and the relationship that I had with Baryalay, as well as with all my other informants, there is the important difference that *I* was the one who needed to talk to them, the one who "went" to them for information, and not vice versa. There was no pretense on my part of holding any "healing" goal in our sessions, nor were my informants explicitly speaking to me in order to gain any amelioration in their psychological condition. Also, the flow of the material reward (the money) for what was being done, in the case of Baryalay, was going in the opposite direction from the one it usually goes in a psychotherapeutic dyad (there was no money involved with any other informant). In spite of all the differences, however, several of the mechanisms that are usually triggered in a psychodynamic encounter did present themselves during my sessions with my informants, particularly with Baryalay. Free association often took place, clues of transferential material emerged from our conversations, as surely did countertransferential material on my part. Likewise, despite the fact that no "healing" goal was intended in the work with my informants, nevertheless I did detect a very clear sense of relief on their part while talking to me about their private issues and predicaments. While I certainly embodied the odd and nosy stranger, the outsider that had no part in their private (and social) lives, for the very same reason I represented a safe repository for the content of whatever their conversations with me would ever disclose. In a sociocultural environment where privacy is a very hard feature to attain and maintain, and where even your best friend or your closest brother will not (very often) keep to themselves details of your personal life that would better not become public, the possibility to open up to someone who will not be able (even if he wanted to) to pass your private information to anybody related to you, certainly must have been perceived as an asset in my role, which played to my advantage. And indeed, I did feel that there was sometimes much relief in the way my informants told me about their problems or private emotions. More than once I was told that a certain detail was being recounted to me for the first time in their lives, which happened with a sense of "liberation" in my informants, very clear to my eyes.

A family portrait

When our interview sessions began, Baryalay had been married for about three years and had a little daughter who was almost two years old. He was obviously very happy about his daughter, and must have been an affectionate father, from what I could gather during our meetings and the many phone conversations he had with his wife while I was present. He was born in the village where he is still living. He lived in the village until he started elementary school, at which point he moved to Peshawar (around 1989). He finished first grade there, then returned with his family to the village, where he completed elementary school. Subsequently, his family moved again to Peshawar (circa 1995), but they remained there only one year. Upon returning to the village with his family, Baryalay was enrolled in Nangarhar High School in Jalalabad (the main state-run school in the city), from which he graduated. In 2003 he enrolled in Khost University. Living in Peshawar during his childhood and early adolescence, albeit intermittently, and going there sporadically to see family members in later years, must have been in itself an experience that marked him deeply. The moral exceptionalism that he attributes to himself and his family (with the exception of his father, as we will see) vis-à-vis the rest of the village community, pivots also, albeit not exclusively, around the "cosmopolitan" experiences that the family enjoyed in Peshawar. During one of our sessions, the very mention of his years of childhood and adolescence came as an associative thought while talking about interpersonal conflicts among villagers. In explaining the reasons why village life is so rife with jealousies and interpersonal frictions, Baryalay made a point of distancing himself from the rest of his fellow villagers.

> I am not like these people [the inhabitants of his village]. I have seen other places, I have lived in a different way. I have gone to university far from home, and met people from all over. Many of these people have never left their village, they don't know how to live properly. They are ignorant and uneducated. That's why they are constantly fighting against each other, backbiting . . . they don't have trust in each other, in anyone. . . . I don't like this kind of life, to me it doesn't make sense.

Baryalay chose not to talk very much about his natal family, and I did not push him to do so. From the few details that he gave me about it, I perceived a degree of embarrassment and, perhaps, shame at what had happened in his family because of his father's lifestyle (about which, see below). He has two older brothers and one younger brother. His mother died when he was approximately five years old. He vaguely remembers someone coming to him, while he was in bed, to tell him that his mother had passed away. His attitude toward these memories is detached and almost indifferent. At that time, his father was approximately fifty years old and decided immediately to find another wife for himself. He chose a twenty-year-old non-pacha girl from the outskirts of Jalalabad, whom he married one year later.

The family of the girl was happy to give her to my father. You know, at that time we were in a good economic situation. My father had his carpentry business, and his small construction company as well. It looked like a good investment to give their daughter to an old man who had money. My family, however, did not like the idea. My brothers [his two older brothers, at that time twenty-two and twenty-five years old] *tried to convince my father not to marry again, and not to have other children. They said that it was irresponsible of him to have more children, because if he had suddenly died, while the children were still small, it would have been up to them to take care of their stepbrothers. But my father responded with arrogance* [ghoror] *and insolence . . . he said that he would decide himself what to do with his life, and that nobody could tell him what to do . . . so after a while he did start to have children with the new wife. In the end he had three girls and three boys . . . and he is still alive.*

ANDREA—*How is your relationship with your father now?*

BARYALAY—*People in my family are afraid of my father. He is illiterate, he is violent* [zurawaar], *he lives like the other people in the village, he lives by the same rules. He went to live with his new wife and his children in another house in the village. We don't meet very often, we live separate lives now. We see each other only for* ghamuna *and* khoshaluna [the joyful and sorrowful occasions in the life of a family, mostly weddings and funerals, when relatives from the same lineage gather together].

ANDREA—*Are you angry at your father?*

BARYALAY—*No, I guess not. . . . I don't care anymore, I just don't consider him. . . . I have no interest in having a relationship with him at this point.*

Baryalay takes much pride in his role of sayed, or pacha. It is an important component of his self-representation. He believes that he is part of a genealogy that can be traced back to the Prophet Mohammed (albeit not in literal terms: he would not be able to name all his predecessors up to the Prophet, of course), and he feels deeply the responsibilities that come with such role. For this reason, I perceived that he considers the behavior that his father displayed throughout his life as being shamefully not adherent to the values that his family should embody: piety, wisdom, detachment from what he considers the average Pashtun man's ethical flaws—such as proneness to violence and disregard of Qur'anic principles to the advantage of customary norms (*rawaj*)—all virtues that, in Baryalay's opinion, his father did not possess. Yet the disdain that Baryalay feels toward his father, which surfaced in our conversations, is also tinged by the deep sense of respect and deference to the paternal authority that, in Pashtun culture, inescapably informs the relationship between father and sons. To this, we must also add fear, about which Baryalay talks in collective terms ("People in my family are afraid of my father"), but to which he himself is also obviously subject. The inner conflict that Baryalay perceives between the culturally required respect and deference that he is supposed to grant his father, and the private revulsion that he feels toward the way his father behaves and publicly presents himself, is, in my opinion, reflected in the scarcity of occasions in which he mentions his father at all. In one of the other few conversations wherein this happened, Baryalay was explaining to me why he prefers not to participate in, and get carried away by, the many instances of interpersonal and interfamilial conflict that often arise in the village (which we will detail better below). He associated interestingly at the end of the following passage:

> *My family members and I do not fight in the village. We don't do these things. We try always to solve problems through talking, and, if necessary, with a Jirga* [the council of elders that customarily may be summoned to avoid violent conflict and solve a dispute]. *The uneducated and ignorant people in the village call us* beghairata [unmanly, in this context similarly to the Western colloquial expression "without balls"], *because to them we cannot fight to defend our rights and what belongs to us. But others in the village, those who are more civilized and educated, they understand that we are doing the right thing, they consider us well because we try to solve things peacefully.* . . . [long pause] *The only person in my family who fights and uses violence is my father.* . . . *He calls us* beghairata *as well, like the others do. He says we are* beghairata *and shameful* [besharm].

The rejection and disgust that his father openly expresses at Baryalay's life choices, as well as at the public persona that he has elected to embody, clearly hurt Baryalay, as I interpret this last passage to indicate. Nevertheless, the moral role that his position as a pacha culturally enjoins upon him remains crucial for the construction of a meaningful self.

Baryalay seems to suffer no hesitation in this regard. Notwithstanding the pain of being rejected by his father, Baryalay's condemnation of the latter's lifestyle and public behavior seems to leave no room for rethinking. Indeed, his two older brothers appear to have chosen the same path of interaction toward their father (the youngest brother is attending university in Jalalabad). His next-oldest brother is forty-seven years old and lives in Peshawar. He is a mullah and works for a local mosque in the city. Baryalay has always remarked about him that he is an "open-minded" mullah, who left the village (and Afghanistan altogether, as a matter of fact) in order to free himself from a lifestyle he did not approve of, nor enjoyed. Baryalay's oldest brother is fifty years old and lives in the village with him, in the house that Baryalay occupies with his wife and daughter. He is only partially educated, and owns a shop in the bazaar of the village. He is married, but cannot have children. Their mullah brother allowed one of his two children to be adopted by his shopkeeper brother in the village. Baryalay recounted that the latter and his wife were desperate after realizing they could not have children, and the adoption of their nephew was received with relief by the whole family. The relationship between Baryalay and his three brothers is close and cooperative. Baryalay visits his mullah brother in Peshawar on a regular basis, and he comes to visit Baryalay to the village as well. They both financially support their shopkeeper brother in the village, who apparently often does not manage to make ends meet. All three are keeping the youngest brother in university, while also preparing for the big expenses that his future wedding ceremony and interfamilial economic arrangements will require.

> *My father has spent most of his life in the village, together with the other villagers. He has learned from them . . . he does not behave like a pacha, he behaves like one of them. My brothers, sisters and I have lived in other places too, we have lived with other relatives who have taught us what it means to be a pacha, and how it is appropriate that a pacha behave. When I was in Peshawar with my brother* [the mullah], *he took care of my education. In the village, my brother-in-law* [husband of his older sister] *showed me how to behave like a pacha.*
>
> ANDREA—*What about your mother? Did she educate you in the pacha tradition?*
>
> BARYALAY—*No, my mother was not a pacha, she did not know anything about these things. She was a simple woman. . . .*

Finding an alternative role

Thus, while Baryalay's father was present throughout his childhood and
adolescence, the example and role model that he offered to Baryalay and
his brothers were rejected by them as negative, undesirable, and morally unacceptable. Other surrogate father-figures proved more valuable for
Baryalay: his older mullah brother, other male members of his extended
family, as well as members of the in-law lineages that joined his familial environment. The rejection that Baryalay and his brothers have demonstrated
against their father seems to rely mainly on cultural bases. The context
wherein Baryalay and his relatives grew up was strongly informed by the sentiment of moral exceptionalism that is peculiar to the pacha environment.
As a social group that claims direct descent from the Prophet Mohammed,
certain diacritica in ethical behavior and moral values are considered to be
crucial to one's public persona, as well as private self-representation. The
open disregard for such ethics and values on the part of Baryalay's father
put him outside the boundaries of the social group to which he belonged
by ascription (see Barth 1969). The self-inflicted social ostracism (from his
own pacha group) that Baryalay's father chose to subject himself to operated
evidently with a greater force than the emotional and primary ties that his
sons must nevertheless have had with him. In other words, the gravity of the
cultural breach that the father carried out publicly must have overcome the
sentiments of filial piety that his sons surely must have privately harbored and
perceived. The construction of a self that would fully reflect the requirements
of a culturally appropriate pacha man turned out to be more important for
Baryalay and his brothers. This certainly has something to do with the social
narrative that the *pachaiaan* (pl. of pacha) have construed for themselves,
as a sort of "special" sub-ethnic group with an ambiguous relationship with
the broader ethnic group to which they also proudly belong—the Pashtuns.
Baryalay is adamant about the elitist sentiment that the pachaiaan hold for
themselves as opposed to the "average," the "other" Pashtuns. From a utilitarian, if unconscious, viewpoint, it pays a relevant dividend to maintain
one's identity as pacha well defined and even ahead of being Pashtuns.

BARYALAY—*We are pachaiaan, we descend from the Prophet. . . .*
ANDREA—*But you are Pashtuns. . . .*
BARYALAY—*Yes, we are Pashtuns* and *pachaiaan. . . .*
ANDREA—*Do the other people in the village consider you Pashtuns as well?*

BARYALAY—*Yes, they too think that we are both Pashtun and pachaiaan.*

ANDREA—*Do you know where your family roots are in the Middle East?*

BARYALAY—*No, I do not know.*

ANDREA—*So, what does being a pacha mean to you?*

BARYALAY—*We know that we descend from the Prophet, and being good Muslims for us is more important than being good Pashtuns. We are educated in considering the Qur'an more important than anything else, for instance* rawaj [customs]. *We know the Qur'an, we read it and we study it. I know that there are many parts of* pukhto *that are not in line with what the Qur'an says. We do not follow those customs that are in conflict with the Qur'an. The other villagers* [i.e., the "average" Pashtuns] *do not care. They are ignorant, they do not know the Qur'an, and for them rawaj is the same thing as the Qur'an. But we know better, and refuse to do certain things that are against the Qur'an.*

ANDREA—*Like what?*

BARYALAY—*Well, for example* walwar [brideprice]: *we do not ask for walwar from the family of the groom. We only pay* mahar [i.e., a quantity in gold to the bride, which in theory belongs directly to the bride, and not to her father, as the walwar would] . . . *mahar is in the Qur'an, so we pay mahar . . . walwar is against Islam. People here go bankrupt because of the walwar, also poor people have to pay a lot of money if they want to get married . . . this is not right. Also, we keep our weddings modest, so that people do not waste too much money on something that is only for show. People here* [the "other" Pashtuns] *have a lot of jealousy* [bakhiltob] *when it comes to weddings . . . they have to show they are better than their neighbors, and so for arrogance* [ghoror] *they waste a lot of money that should be spent doing something better for their families. This is not right. We do not do these things . . .* [long pause] *We do not like to fight like the others do. We try to solve problems in a peaceful way. We do not like violence. This kind of life that they live, always fighting, and quarreling, and being jealous of each other, this is against Islam.*

Although Baryalay does not say it explicitly, it is apparent that he considers pachaiaan to be "better" than their "average" Pashtun co-villagers. The preeminence that pachaiaan accord to the tenets of the Qur'an, to the disadvantage of the principles of pukhto, clearly makes the pachaiaan morally superior to the other Pashtuns, in Baryalay's opinion. The relationship that pachaiaan in his village (and elsewhere in Pashtun areas) have with the rest of the Pashtuns living alongside them is ambivalent, to say the least. Baryalay claims that his co-villagers (who belong mainly to two other Pashtun groups: Mohmand and Mangal, the latter having immigrated from Paktia province) consider them to be Pashtun, like everybody else. In different rural settings throughout the province, however, I found that often the pachaiaan (or *saydaan*, as they are also known in Nangarhar province) are viewed in a slightly different light vis-à-vis their co-villagers. In one specific case, among the Shinwari group, close to the border with Pakistan, one family was clearly described to me as having come to the area about 150 years earlier from Iraq. They were believed to be descendants of the Prophet and Pashtuns at the same time, as the pachaiaan of Baryalay's family are, and yet the memory of their Middle Eastern provenance was firmly kept by their Pashtun Shinwaris co-villagers, who sometimes used their "non-pure" Pashtun-ness to justify resentments and rancors that owed more to economic reasons than ethnic ones.

In fact, pachaiaan and saydaan have always been respected and revered as individuals who may have a closer connection to Allah, in light of their presumed descent from the Prophet. Paying homage to a pacha or sayed is still believed by many to bring about rewards in the afterlife. Such acts of deference often take the shape of donations, usually in kind. Not infrequently, pachaiaan are bequeathed pieces of property or land by deceased individuals in their wills. Whereas these acts are felt by many as necessary to improve the chances of a better standing after death, others (and even sometimes the same who perform these very acts, in an almost contradictory manner) resent bitterly the pachaiaan's accumulation of properties and riches accomplished without the supposedly appropriate labor and risk-taking that would be required of a non-pacha Pashtun in order to reach the same degree of affluence. Baryalay's family, by his own admission, is no exception in this regard, and is in fact the most affluent in the village. In an ecological environment where land is scarce, and each family (when owning any land at all) possesses an average of about 1 to 2 jiribs (two and half jiribs are equivalent to one acre), a property of about 15 jiribs, as Baryalay's family has, means faring well above the average. As it often happens, Baryalay's family's land is farmed by hired laborers, who do not possess any land of their own, because landowners generally see working the land themselves as socially demeaning.[1]

---

[1] In their classic case studies of Swat, Pakistan, both Fredrik Barth and Charles Lindholm (Barth 1959, Lindholm 1982) described the sayyeds within the Pashtun environment they studied as definitely "non-Pashtun." In other words, it was assumed, and accepted, by the locals and the sayyeds themselves, that they did not belong to the Pashtun "aristocracy" that was politically and

They do not kiss our hands any longer [as a sign of respect] . . . *this happened in my grandfather's time, but now nobody does it anymore. They do respect us for what we are, however. They know we are pachaiaan, they know what it means. We do things differently from them . . . we always keep in mind the Qur'an and we behave accordingly . . . we are expected to behave in a different way . . .* [long pause] *I do not like the people from the village, they are ignorant, uneducated, and do things that are against the Qur'an . . . we try to stay away from them as much as possible, we do not have many relationship with them. . . . I, for example, pray at home, on my own, I never go to the village mosque, and so do my relatives. . . .*

ANDREA—*What do the other villagers think about it?*

BARYALAY—*They criticize us for this* [peghor warkawi], *they think we should go to the public mosque like everybody else. We also keep our family's women more at home than the other families do . . . they have less relationships with other women than the other villagers have. . . . It's a matter of modesty, of being closer to the Qur'an. . . . However, when they go out, our women are dressed in a different way . . . they do not wear the burqa as the other women do, they wear a hejab or a niqab . . . the burqa is not in the Qur'an, it does not exist in Islam . . . they also criticize us for this, they say we are not doing pukhto the right way. . . . Also the fights* [lanjay] *. . . we stay away from those. There are some villagers who understand that we behave differently because we want to follow the Qur'an, and they respect that. But others, they think that doing pukhto is like following the Qur'an, and they criticize us for that . . . like my father. They say we do not stand up for our rights as a Pashtun should do. . . .*

Baryalay's family (with the exception of his father) is caught in an awkward paradox: as a pacha Pashtun family they are largely expected to uphold a distinctive conduct and behavior, more "Islamic" and pious, as it were. At the same time, however, when this very same pious demeanor diverges from

---

economically dominant in Swat. By being external to any Pashtun lineage, the sayyeds managed to gain for themselves a role as mediators/peacemakers during intracommunity conflicts. In eastern Afghanistan, where scarcity of resources and paucity of available land has impeded the deep social stratification that is found in Swat (and southern Afghanistan), the neat ethnic differentiation between sayyeds and Pashtuns is absent, and the boundaries between the two categories are often blurred, when not altogether nonexistent. We will see in this and the following chapter that often sayyeds become seamlessly incorporated in the Pashtun community, although they might still benefit from the common awareness of their being the "descendants of the Prophet." Baryalay's family chose to emphasize and stand by this aspect of their family history, while his father chose to reject it.

what most consider being an appropriate and required behavioral choice for a "real Pashtun," Baryalay's family members meet the reproach and contempt of many from their village. In other words, precisely those ethical diacritica that define them as pacha (as opposed to "average" Pashtuns), and to which they owe their respected social standing in the community, are at times the source of a loss of social capital. The elitist aspect of this dynamic should not be underestimated, however, and certainly does not go unnoticed by Baryalay's co-villagers. There exists just as much contempt on the part of the villagers against Baryalay's family as there is on the part of Baryalay's family against the villagers because of their inability to conform to those Islamic norms that Baryalay and his family see as patently contravened by Pashtun traditional customs, which the villagers abide by. Baryalay and his family have decided to remove themselves from the social context of the village, except for the basic and necessary relationships. Baryalay admits that the villagers would like his family to participate in the religious rituals of the village alongside its inhabitants. Yet they abstain, in a fashion that is probably seen by the villagers as offensive and "snobby." Furthermore, it must be kept in mind that, customarily, pacha families intermarry with no other group than pachaiaan, which likely increases the perception of a voluntary, presumptuous isolation that the pachaiaan want to pursue vis-à-vis the "average" Pashtun people. Such a kind of quasi-aristocratic attitude that the pachaiaan maintain is all the more problematic because it is based on a religious premise, which would be impossible to contest. The direct genealogical connection to the Prophet, albeit located in a distant past, is accepted as a fact by most (or is at least not questioned by anybody), and it would be unthinkable to refuse (overtly) a higher degree of respect and reverence to someone holding such "pedigree." Openly disrespecting a recognized pacha or sayed would equate to disrespecting the Prophet himself. At the same time, however, within a social environment where personal honor and in-group cohesiveness and loyalty are considered crucial features of any Pashtun individual (both male and female), the defiant self-exclusion that Baryalay's family seems to have pursued on the basis of their perceived moral superiority nonetheless attracts the resentment of the rest of the villagers. The latter likely perceive as "unfair" the use of religious prerogatives that Baryalay's family makes in order to underscore a moral superiority vis-à-vis the rest of the villagers.

Baryalay is certainly aware of the ambivalence of the position of himself and his family in the context of village social dynamics, and claims to have been proactive in the past toward his co-villagers. He recounted how on three different occasions he tried to spur the village notables to get together in the mosque to discuss the status of social relationships in the village, and to find the means to improve a way of life that he deemed too conflictive and acrimonious, and too removed from an Islamic ideal. Only a few of the villagers, Baryalay told me with an aggravated tone of voice, showed interest in his proposal, while most of them dismissed the idea as useless and presumptuous.

> *They told me that they were perfectly ok with the life they were living, that there were no problems in the village. They told me that they did not think that there was too much quarreling and fighting in the village, and that they were good Muslims too, as I was, and that they were as respectful of Islam as I was. They told me that meeting together in the mosque to discuss these things would be a waste of time.*

The frustration of his act of "good faith" only damaged further the opinion that Baryalay held of his co-villagers, and probably sealed the fate of his relationships with them. A complete isolation from the affairs of the village followed the debacle. The ambiguity of the social position that Baryalay maintains among his peers in the village, by virtue of his ascriptive belonging to a lineage connected by blood to the Prophet, is evident in the way he describes the responses of the villagers to his proposal. After all, he is not formally trained in religious studies, and technically he holds no higher qualification in religious matters than any other villager. The fact that he is the village's only university graduate (and in agronomy, at that) does not win him enough clout in this case. Whether they really liked their lives as they were or not, the villagers were evidently piqued by Baryalay's not-so-veiled suggestion that not only their conduct, which they considered in line with doing pukhto, was disruptive of social harmony, but also, and worse, that their social behavior fell short of what was morally enjoined on them by the Qur'an. In Baryalay's village, it is apparent that pachaiaan have to walk a socially thin line, and administer wisely the prerogatives that their status provides them with.[2] They have to attentively manage their identities as both Pashtuns and putative descendants from the Prophet.

---

[2] Such prerogatives, as we have partially seen, revolve generally around the "respect" that others are forced to grant them on the basis of their saintly lineage. This can materialize in, among other things, stronger political ascendant among state bureaucrats, "gifts" in land or other kinds of goods from co-villagers or acquaintances, or being considered as a particularly influential and capable individual during the negotiations to solve an intracommunity conflict.

The father of Baryalay, in this regard, chose a radically univocal position for himself. He vocally embraced his being first and foremost a Pashtun and, to the dismay of his family members, aligned himself with those whom Baryalay considers ignorant and violent individuals, although their behavior meets the ethical requirements expected in a "real Pashtun." According to Baryalay, his father, therefore, has no problems of adjustment to the village life, and does not suffer the criticism of his co-villagers. The price his father paid was the loss of consideration and respect by his own group that no longer considers him a pacha. We do not know whether, in the eyes of the villagers, it might be just Baryalay's father who embodies the perfect features of a "legitimate" Pashtun-cum-pacha, holding full membership in both groups. At any rate, the fact that Baryalay keeps associating the image of his father with that of the contemptuous villagers speaks to the deep discomfort that his strained relations with his father must cause in him, albeit only seldom verbally articulated.

Shaping one's own masculinity

During our sessions, Baryalay elaborated often on his life in the village and the personal frustration that it entails.

There is a lot of enmity [dukhmani] *and jealousy* [bakhiltob] *in the village, among the inhabitants. They are always talking bad about each other behind each other's back, they are jealous of what the others have. There is a lot of aggressiveness . . . you have to behave in a certain way, because otherwise the others will consider you weak* [kamzur] *and unmanly* [beghairata], *and will take advantage of you. I can't stand it* [za na sham kawalay che ye wmanam]. . . . *I hate it* [zma bad ye razi].
ANDREA—*Are you the only one with these feelings about life in the village?*
BARYALAY—*No, certainly not. There are other people who do not like it. It is not easy to live like that. There are people who are not cut out for it, who are not made for it. But they have to behave like that anyway . . . it's a matter of survival . . . you have to be arrogant* [mughoror] *and stand up to the others, or else they will be all over you. You have to defend yourself. I know people in the village who are not happy with their lives. . . . Some of them have left the village, they have moved to the city. But not everybody can . . . some have to stay. . . .*

ANDREA—*What about you? Would you like to leave?*

BARYALAY—*Yes, of course, I dream of going to live in Jalalabad with my wife and daughter, but I can't right now. My brother* [the shopkeeper, his oldest brother] *cannot go on without me, I have to help him . . and I do not yet have money enough to live on my own in the city, it's expensive. . . . * [long pause] *Nobody has friends in the village . . . they are always very respectful of each other, they treat each other well, but it's only the outside, they all have black hearts* [tor zra, meaning that they are insincere in their outward expressions].

All things considered, it seems that Baryalay's personal situation in the village is probably less dramatic than that of some others. As a member of the pacha social group, Baryalay finds himself less shackled by the norms and rules of customary Pashtun life in comparison to his non-pacha co-villagers. His social position as pacha allows him to pursue a slightly different kind of life, and legitimately reject some of the constraints that a non-pacha would be helplessly subject to. He can take advantage of the expectations that his social group has to fulfill (in terms of social conduct) in order to shy away from the most odious (to him) aspects of what is generally considered appropriate Pashtun behavior, such as aggressiveness and proneness to violence in achieving ambitions of self-aggrandizement. A culturally accepted alternative avenue for comportment and demeanor is thus available for a pacha like Baryalay. We have already seen that this does not spare him the criticism (*peghor*) of some of his co-villagers, and that for this reason he must navigate the ambiguity and paradox that his position puts him through. Nevertheless, as opposed to the more rigid and uncompromising situation that many co-villagers find themselves trapped into, Baryalay possesses a broader latitude for expressing and materializing his temperamental inclinations, his emotions, and attending behavioral choices. For all the frustration and discomfort that he undoubtedly suffers, Baryalay still has a culturally allowed way out from the worst of the village life, a way to legitimately "rebel" against the most unacceptable aspects of it, which he enacts by removing himself from much of the social life of the village.[3] He pays for this choice with the

---

[3] This is not what his ancestors did (his father being in a particular position within this picture). It emerged clearly during my conversations with Baryalay that his grandfather and great-grandfather (about whom he heard stories when he was growing up) were immersed in their role as pachaiaan in full. They were active in making amulets (mostly pieces of paper with Qur'anic verses written on them, which people would wear as bracelets or necklaces, or keep with themselves), advising people on personal matters, acting as mediators in feuds or conflicts. For these services they received "gifts"

reproach of some of his co-villagers. Yet it is also evident that his position as pacha mitigates the detrimental social consequences of such attitude. Those non-pacha villagers who, like Baryalay, feel that the lifestyle of the village strongly contrasts with how they would in fact like to conduct their lives, cannot afford the luxury to make the same "isolationist" and nonconformist choice that Baryalay and his family have made. If they did, they would condemn themselves to a dishonorable status and a consequent unsustainable social position. They necessarily have to do violence to themselves and adjust their behavior to what they are expected to publicly show. This is obviously the source of deep anxiety, frustration, and exasperation, which often find an outlet in overcompensating behaviors (to use the concept that Edward Sapir introduced upon the suggestion of Alfred Adler. Sapir 2002) in a sort of self-reinforcing vicious circle. Baryalay manages to partially avoid these dynamics. He says:

> *There are people who behave like they would not want to . . . they have to do certain things, but they are not happy with that. So what happens sometimes is that they exaggerate things, they do worse things than others because they know that everyone is looking at them, they want to show everybody that they are good Pashtuns. . . . I know people who do not like dukhmani and lanjay* [enmity and fights], *but who pick up fights more often than others . . . it's crazy* [da liwantob de] *. . . they feel more pressure* [preshaar, an English loan-word] *and exaggerate everything. . . .*
>
> ANDREA—*Does this happen to you as well?*
>
> BARYALAY—*No, I don't do that, I stay away from fights and jealousies. I told you already, my family does not behave like that, this is not what a good Muslim should do. . . .*
>
> ANDREA—*But don't you feel uncomfortable* [marahata] *living like this?*
>
> BARYALAY—*Well, yes, I am under a lot of pressure too . . . it's not nice when you know that people call you beghairata behind your back, I don't like it. I get*

in return. Most of Baryalay's family's landed properties were acquired in such way. They apparently did not eschew social relationships in the village, as Baryalay's family members do now. According to Baryalay, his family's choice of cutting relationships with the co-villagers is a direct outcome of the deterioration that public mores and behaviors were subject to over the three decades of war. As a family who still feels strongly its role as repository of a saintly heritage, revulsion at the current situation was too strong to allow any compromise. Baryalay's father, on the other hand, has apparently endorsed the shift toward the new ethical standards and disavowed his pacha background. There is not much left for Baryalay and his family members to perform as pachaiaan, as their ancestors did within the village previously.

nervous, tense, and sometimes I become upset with my wife or daughter for nothing. . . . I don't like being in the village, and sometimes I react badly with my wife and daughter. . . . I have less patience [sabar] *with them sometimes. . . . I know it's wrong, I feel bad about it. But I am not ok in the village, I feel as if I was in jail, I cannot live the life that I would like . . . and there is also a lot of pressure for the security problems. . . . I am worried, scared, always thinking that something might happen to me or my family* [because of his job—see details below] *. . . it's bad. . . .* [long pause] *Also the people who live in the village, they too are in this situation of constant tension because of the security problems in the district . . . that is also why they behave like that, so suspicious of each other, not being able to make friends, always quarreling with each other . . . everybody is under pressure. . . . Your life is always in doubt* [shak], *there is nothing certain* [mutmaian]. *Today you are rich, tomorrow you are poor. This evening you are alive, tomorrow morning you die. Nobody is certain about anything, they feel that they have to take all that they can right now, because tomorrow who knows what can happen . . . they want to get ahead of all the others* [de tolo na makhke larshi]. . . .

Baryalay manages to escape a psychological dynamic such as behavioral overcompensation thanks to the leeway accorded to him by his particular social position as pacha (and an educated and well-to-do pacha, at that). Nevertheless, he recognizes this dynamic in some of his co-villagers, whom he evidently knows well after all, in spite of his self-inflicted ostracism from the social life of the village. This phenomenon must be so conspicuous to catch his eye.

It is interesting that Baryalay is able to detect as one of the causes for the toxicity of village life the constant situation of tension, danger, and uncertainty that the war and foreign occupation entail at the moment in Afghanistan. On the one hand, he is finding a legitimate justification, based on environmental factors, to explain some of the reproachable (in his opinion) aspects of his fellow villagers' social behavior, redeeming them from a simplistic, sweeping moralistic judgment (e.g., "they are intrinsically and hopelessly bad people," as some members of other ethnic groups in Afghanistan often characterize Pashtuns). On the other, he is building one of the few common grounds that he may possibly share with his co-villagers: the common, profound discomfort and pernicious psychological ramifications that the more than thirty-year-long conflict has had on the lives and social fabric of Afghans. Such common ground, and the feeling of sharing with them an unfortunate fate, allows him (if unconsciously) to relieve himself at least in small part from the otherwise crippling antithesis between his lifestyle and that of the villagers. Certainly his reactions to the difficulties and pressures of everyday life in the village (and in the province as a whole) are different from those that the villagers display—and in that lies the sense of "moral exceptionalism" that he maintains for himself and his family. Yet, he undoubtedly suffers from those feelings and emotions of despair and precariousness that he attributes openly only to his co-villagers, just as much as they do. Paradoxically, however, some of his co-villagers represent just the most proximate cause for part of his problems with security in the village and Jalalabad. His job as employee of a foreign-funded agricultural NGO puts him in an uncomfortable and dangerous condition in the context of his home district, which, in the years prior to my work in Nangarhar province, and increasingly during my fieldwork research there, became a hotbed for antigovernment activities and insurgency. In a volatile and overall "anxious" environment such as that of rural Pashtun Afghanistan in 2012 and 2013, it takes very little for anybody (including one's close relatives) to be branded as a *kafir* (infidel), a *jusus* (spy), or an enemy in general (*dukhmaan*), with all the attending negative consequences (often deadly). Baryalay explains:

Half of the people in my village are ok with me. They are the farmers. They roughly
   know what I do for a living, I help them with their fields, and they are happy
   with the support I give them. Not that I can bring the projects of the NGO to the
   village—that would expose me too much—but I help them nonetheless with
   advice and know-how. The others, the ones who don't like me, are those who
   either work in other areas than agriculture, or are unemployed. They don't
   know for sure where I work [he keeps the details of his occupation to himself], *but they know I work in Jalalabad, with projects in agriculture . . . they*
   imagine that I probably work either for the government or for a foreign NGO.
   They look at me with suspicion. They hate whoever works for the government or for the angrez [foreigners]. *I know that someone knows what I do for*
   work . . . nothing is really a secret [raaz] *in the village. That's why I try to go*
   outside of the house as little as possible when I am in the village. As soon as
   the sun goes down, I get inside and don't go out until morning. [long pause]
   These jobless young men . . . they don't go to school, they look for a job, don't
   find it, end up with having nothing to do all day, and they get picked up by the
   Wahhabis, the preachers. . . . These young men are angry [khapa], *because they*

are jobless, they think it is because of the state, because of the foreigners who run the country for the Afghan people . . . they swallow all the propaganda that these Wahhabi preachers feed them . . . for them [the preachers] everybody is a kafir . . . government workers, NGO workers, all the foreigners, everybody. [The preachers think that] kafirs who are in Afghanistan have to be expelled from the country or killed . . . they don't make compromises . . . they are crazy. These preachers come from Pakistan when the snow melts in the mountains, bringing money and propaganda, then they go back before the winter, but they leave new [Afghan] followers here . . . those jobless kids are angry, hopeless, they listen to them [the preachers], they follow them . . . they don't understand. The preachers give them money, they help their families. . . . But nobody knows for sure who is working for the insurgents [yaghiaan, literally "rebels"] or the Taliban. They have a normal life, with whatever they do during the day. If there is an operation, they participate, then they go back to their normal occupations. Nobody knows for sure who they are. Even when you see the Taliban on the roads or in the fields, they have their faces covered, you don't know who they are . . . it could be your neighbor. It's dangerous to talk to anybody . . . if you say to somebody that you think the Taliban are not right, and that person works for them, you're in trouble. If you admit to the wrong person to fearing for your life, that will be taken as a proof that you have something to hide, and you're in trouble. If somebody tells you that he is a member of the Taliban, and you ask why he chose to become one, you are in trouble. It's horrible, you can trust no one, you cannot speak to anybody. Every day is like this. . . . I can't stand it anymore [nur na sham kawalay che da wmanam]. . . .

Thus, notwithstanding the (socially) "shielded" condition that Baryalay and his family enjoy in the village, the general situation of daily life that everyone is subject to there, with its stresses, anxieties, and uncertainties, affects Baryalay heavily nonetheless. He feels cornered and constantly in danger. The atmosphere in the village seems toxic for several reasons. "Nothing is really a secret in the village," Baryalay previously stated: as in most small communities, regardless of the cultural context, villagers talk about each other, and take pleasure at the others' missteps. We have encountered this situation also in the accounts given by Rohullah and Umar, whereby no one, not even one's own brother or best friend, can be fully trusted in keeping personal or sensitive information away from the gaze of others. Baryalay's job puts him at risk with the many (increasing day by day) who consider anybody working for the government or the foreigners as an enemy. The reality that antigovernment elements maintain in many areas of the country is often based on an uncompromising us-them dichotomy, the good against the evil, the Islamic against the un-Islamic. There are few subtleties in such narrative, very few nuances that a man can count on. The slightest doubt or suspicion is immediately turned into a reality that requires retribution. Those who, like Baryalay, do not buy into the rhetoric of the radical Islamic preachers, and, by necessity or conviction, work for government or foreign institutions, live in terror of possible negative consequences.

During the first months of our relationship, Baryalay arranged to come to Jalalabad on purpose to have class with me (his job as a field officer required him to work in the districts, and he did not have to join his colleagues in the Jalalabad office very often). He used to ride a motorcycle with a friend of his from Jalalabad to the village, and always took care to cover his face with a scarf during the trip, in order to avoid anybody recognizing him. When the situation in the district he went through on the way to Jalalabad worsened over time, he decided to start making the trip back and forth by shared taxi, which, in his opinion, ensured a better anonymity along the way. Even so, the cars he traveled in were stopped several times and searched by Taliban members at improvised illegal checkpoints. He never had on himself any piece of documentation that could betray his collaboration with a foreign NGO, and never really got in danger. Yet the line is extremely thin, and disaster could happen any time. Midway through our classes, he was reassigned to a desk job at the NGO headquarters in Jalalabad. He had to rent a room in the city, which was problematic with his salary, and started to go home only for the weekend. Whereas the new arrangement allowed him for more peace of mind, the forced separation from his wife and daughter felt painful to him. He had no alternative, however. He was the only real provider for his family and that of his shopkeeper brother, who could not make ends meet (the revenues from the fields his family owned went entirely still to his father, who retained exclusive ownership over them).[4] Thus, although there was a good share of villagers who did not dislike him, due to the technical help that he managed to give them in the fields, the many others who did not profit

---

[4] As usual among Pashtun families, there is no subdivision of property until the father dies. Only then will the sons receive a portion of the inheritance (daughters receive nothing of the inheritance). All that is legally owned by the family is usually in the name of the father, who is not bound to divide his property in equal parts among his sons in a written will. The ramifications of this state of facts are numerous. The father obviously remains in a strong position (a sort of modern *paterfamilias*), and continues to exert his authority until the end. This situation is sometimes the source of resentment among the sons toward the father, and certainly of competition among the brothers. More important,

from his know-how or professional position saw no reason for not applying to his case the principles and ideological prejudices that the radical preachers had successfully propagated among the community members. In their eyes, Baryalay as well constituted a potential enemy, a kafir, and could be treated as such (i.e., killed, in today's Afghanistan). Every day Baryalay walked a very thin line, trying to control the damage and risks at his best. The confinement to his house, and the truncated relationships with many among the villagers that he resorted to, are due both to his personal impatience with his co-villagers' lifestyle, and to pragmatic choices related to personal security. In this light, the "lenient" and understanding attitude that Baryalay shows toward the "wrong" behavior and political choices of his co-villagers is all the more remarkable ("that is also why they behave like that, so suspicious of each other, not being able to make friends, always quarreling with each other . . . everybody is under pressure. . .").

Baryalay traces his frustration also to something beyond mere personal security: "I am under a lot of pressure too . . . it's not nice when you know that people call you beghairata behind your back, I don't like it. I get nervous, tense." He has elected not to care excessively about the fact that he is considered by some a beghairata. He goes ahead with his personal choices and his lifestyle as a whole because he deems them appropriate to his social status and position as a pacha. However, he is also a Pashtun, and as a Pashtun he has been raised— albeit as a Pashtun-cum-pacha. Baryalay's ideal of *nartob* (manliness, masculinity) reflects an ideal that many share, but that few see implemented in reality among today's Afghan Pashtuns. As Baryalay also suggested, the blame is often placed onto the social devastation and disruption that the war(s) have caused to Pashtun communities. *Nar* is not just a generic man (for which meaning the term *saray* is used). Nar is a manly man, a man with specific attributes peculiar to true respectable Pashtun men. Baryalay explains:

> He who does pukhto [pukhto paali] *is a nar. In my district there lives a big landowner, a khan. He is not the richest of all, but he is wealthy. He is never*

in the case the father should die unexpectedly (which is common) without writing a will, the household might fall into chaos, due to the lack of instructions on what to do with the family property. State law does not hold much weight in these circumstances. It is in these cases that some of the worst intrafamily feuds take place, as I was told. In Baryalay's case, his father is still alive, and owns all the family land. However, due to his resentment toward his father, or his revulsion at the village situation, Baryalay told me more than once that he does not want anything to do with the family's landed properties. "Land brings only problems and fights," he once told me. "I don't want any land, let somebody else have it."

stingy, he always helps others who may be in economic difficulty, always offers
more than his due for communal necessities, shows hospitality [melmastia], *he*
behaves with wisdom and restraint. For example, some time ago the villagers
decided that it was necessary to build a new bridge over the river that flows
near the village. Every villager had to give a determined amount of money
so that the company building the bridge could be paid. Some of the villagers
could not pay their quota because they were too poor, others only faked being
poor and did not pay. The khan gave himself more quotas than he should have
to cover those who could not or did not pay. That's nartob, he is a nar.

Masculinity, nartob, as intended in the ideal model of a Pashtun man, is evidently still relevant for Baryalay, and he does not reject its cultural schema.
From the story of the rich khan and the bridge, we can realize how much
the idea of generosity, of self-sacrifice for the good of the community, is also
an integral part of the idea of masculinity that Baryalay endorses. What he
wholeheartedly rejects is the current interpretation of nartob that he sees
common among many of his peers. Violence, abuse, arrogance, illegitimate
appropriation, is what Baryalay sees nartob as having been turned into by
the thirty-plus years of war. He is well aware that, in order to comprehend
fully the roots and the current aspect of Pashtun cultural models and people's
public behaviors, the perspective has to go beyond the present, and reach
back in time to understand crucial historical contingencies. As Eleanor
Leacock has convincingly suggested in the case of many postcolonial and
semi-colonial sociocultural environments, also for Afghan Pashtun culture
(albeit lacking a true colonial history), ignoring the impact of dramatic past
events, and keeping one's analysis tied to a short-sighted ethnographic present, would be "unscientific and unethical" (Leacock, 1983: 272). So, as it
happened with many of my informants, Baryalay remains strongly attached
to an ideal of nartob that, most likely, has been lost to many Afghan Pashtuns
amid the decades-long conditions of social destabilization. The enculturation
he underwent, and the general environment in which he grew up, had (not
surprisingly) a powerful impact on the construction of his self. His mellow
personality traits, as well as the education into the pacha *Weltanschauung*
and core values he received during childhood and adolescence, had to find
a way to coexist with the imposing presence of pukhto in the community he
lived in. As a man who recognizes himself as a pacha and as a Pashtun alike,
the ideal narrative about the "true Pashtun man" (pukhto) must have resonated forcefully in Baryalay, and must have served as an important source

for self-identification. Albeit purged of its aspects that seem incompatible with the main guidelines of pacha ethics (interestingly enough, embodied by the figure of Baryalay's father), nevertheless pukhto is irrefutably a component of Baryalay's self-representation, as well as the scaffold around which one of his competing (or complementary) selves revolves. Baryalay showed such plurality of selves in recounting one event that took place a few months before our conversation, in the midst of discussing extramarital affairs in his community.

In reality, for both men and women, if they want to, it is possible to have extramarital relations. For example, in a village near my village lives a farmer who has a sexual relationship with a forty-year-old married woman. This woman lives in a different village, also near to my village, and is married to a very old man. They had children in the past, but now he cannot have sex with her anymore, so she found a lover in the farmer. The old man knows about this, but he is too old to do anything himself about it, and their children are still too small to take action [i.e., to punish the woman and her lover for dishonoring the family, presumably by killing both]. The farmer, the lover of this woman, is also married. He does not want to have sex with his wife anymore, and his wife has found herself a lover in her village. You know, she is still young, certain impulses are natural, and cannot go unsatisfied for long. The farmer knows that his wife has a lover, but he does not care. To be honest, everybody in the area knows what is going on in those houses. In my village, we don't like this situation at all. It is bad, it is sinful, it is against Islam and against rawaj [Pashtun custom]. We don't want something like this happening where we live. With some others in my village we have planned to get a hold of the farmer and beat him up, so that he learns his lesson and stops doing these things. If the family of the old man [i.e., his male small children] cannot protect his honor and respectability, we will do it for him. What is happening is not acceptable [de manalo na de]. If the farmer starts having sex with his wife again, also his wife will stop having a lover. We have yet to carry out the plan, because we are worried about the consequences for us . . . we don't want big problems. . . . You see, it is not too difficult to have an illicit sexual affair in the countryside . . . you can meet in many places, in the fields, in a tractor along a small road . . . people do it, but it's wrong, I don't like it. . . .

ANDREA—Well, many of my [male] friends have confessed to having extramarital relationships . . . it seems a pretty common thing. . . .

BARYALAY—*Yes, actually for men it is . . . but it's really wrong. . . . I made a promise to my wife, and I will keep it. It is important to me, and it is against Islam to do otherwise.*

ANDREA—*So, you know that you will never cheat on your wife.*

BARYALAY—*No, I don't want to cheat on my wife, no . . . only if I have to stay away from her for a very long time . . . say one or two years, without ever seeing her, then I think that I would do it, because certain natural things cannot be kept down for too long. . . .*

ANDREA—*I understand . . . however, even in that case, your wife, being at home with someone from your family, will be controlled and will not be able to satisfy her own natural impulses. . . .*

BARYALAY—*Yes I know, you are right, it is not fair . . . women here live like in a prison, they cannot escape easily . . . but they too have their own natural impulses. . . .*

In his account of the intricate story of multiple adulteries, and his reaction to them, we can neatly detect the overlapping of the pacha and Pashtun selves that Baryalay maintains within himself. His religious upbringing makes him consider as unacceptable the sexually improper conduct of the protagonists of this story. Their behavior is un-Islamic and contrary to the main tenets of the Qur'an. As a pacha he cannot gloss over it. As a Pashtun, however, Baryalay believes not only that such behavior brings dishonor to the community at large in which he lives (note that the people involved are not living in his village), but that, more important, the compromised honor of the old man and the whole community must be restored through some sort of retributive (violent) action. In pukhto, the Pashtuns' customs, it is men who take responsibility for redressing wrongs that involve the honor of a family or an individual. Masculinity revolves around the willingness to take action in cases such as this one (that is, it becomes a matter of *ghairat*). Against the practical impossibility for the old man's family to take things into their own hands (the children of the man are too young to do so), dishonor falls upon the community that implicitly allows a such state of facts to continue. This is an interesting case wherein the concept of honor, and its dynamics, is taken to an extreme. By this I mean that the symbolism surrounding honor here does not apply only to the individuals directly involved in the case (i.e., the extended families of the protagonists), but extends also to those who would technically be not involved, and yet are considered to be so by way of "analogy"—metonymically, as it were. The morally reprehensible state in which the individuals directly concerned are living, if not acted upon, will be considered as spreading, like a "contagion," onto the whole community that hosts them. Inaction is considered to be connivance, and responsibility becomes shared, bringing about shame for all. Sure enough, the method that Baryalay and his friends intended to use against the adulterer and/or adulteress, in order to preserve the moral integrity of the community (and discourage the repetition of similar behaviors), falls short of being a capital punishment, which is usually the case when sexual misconduct is involved. In this case, the degree of violence (planned or executed) seems to be inversely proportional to the distance in kin relationships. If the sons of the old man and the adulteress had been able to bear arms, they would have been expected to kill either the adulterer, or their own mother (the adulteress), or preferably both.

Yet such a mechanism of metaphorical "contagion" is at work also in other circumstances, which are beyond local preoccupations with community honorability. We have seen it working in the previous informants' stories, particularly when related to the loss of social respectability for individuals associating with supposed disreputable people, such as foreigners or government officials. In 2012 and 2013, in the midst of a popular revolt against a foreign military occupation and a government that is seen as a lackey of the "infidels," it took very little to fall from grace within a Pashtun rural community. We will see that the suspicion of having been "infected" (to remain within the medical metaphor) by the wrong ideas or moral values is susceptible to be leveled against anybody (including close kin members) for even seemingly trivial "missteps," such as speaking with a foreign person or spending too much time in the city away from the controlling gaze of one's fellow villagers. Once the suspicion has been unleashed, any repercussion is potentially expectable, from simple social ostracism to violent retribution. Yet punishment, in such cases of politically tinged moral misconduct (let us not forget that "jihad" is just as much a religious endeavor as it is a political one, especially in the twenty-first century), comes often from apparatuses external to family and kin. In contemporary, post-9/11 Afghanistan, the politico-religious policing of the community has been taken over by those who claim to be the representative of the "true" Islam and to be fighting for it on occupied soil—the so-called Taliban and their satellite groups. Countless individuals, suspected or accused of (improbable) connivance with Western powers or the Afghan government, have been kidnapped and/or killed during my stay in Afghanistan for precisely these reasons. Baryalay himself is the victim of the mechanism of "contagion" because of his job for a foreign NGO. Regardless of what his ideas, moral values, or behavior really reflect, the possibility that he might be "infected" by an un-Islamic mindset is already open. As we have previously seen, his fears and anxieties spring partly from this situation.[5] Such a phenomenon is slightly different, I believe, from the usual dynamics that obtain in similar cases of occupation by external military forces, wherein usually the "collaborators" are subject to retribution by the community members who manage to uncover their participation in the occupier's policies. Their actions, however, must generally be in material support of the occupier.

In today's Afghan Pashtun context, though, material help to the NATO forces in the country, or the Afghan government, is not an indispensable requisite to be considered a "collaborator," and be punished accordingly. Something more impalpable and "invisible" becomes the catalyst for such a phenomenon in (particularly) rural Pashtun Afghanistan. For this reason the metaphor of the "contagion" seems to me especially apt. There is no way to demonstrate that one has not been "infected." The "disease" does not show, as the providing of material support for the enemy would upon investigation. There is no way in which the individual who is accused of having become "un-Islamic" or a "bad Muslim" because of his (even loose) association with non-Pashtun, non-Afghan individuals, may restore his good name. Such a condition is so ineffable, so "unfalsifiable" (to paraphrase Karl Popper's terms), that, once attached, it stays on almost indefinitely.

The psychological mechanism undergirding this phenomenon has its roots certainly in fear, uncertainty, and the premise of superiority (vis-à-vis all other Afghans) that Pashtuns nurture for themselves. Amid a national narrative, widespread down to the smallest communities in Pashtun areas of the country, that has historically exalted the position of moral privilege and consequent political power that the Pashtun ethnic group held within

[5] About eight months after my final departure from Afghanistan (June 2013), Baryalay updated me by phone about a disturbing event that had taken place a few days earlier. His youngest brother (the university student in Jalalabad) was kidnapped by unknown masked men from a shared taxi on his way back to the village. They stopped the car at an illegal checkpoint, screened the ID cards of all the passengers, and extracted Baryalay's brother from the car. They accused him of working for the NGO for which Baryalay is working. His brother protested that he was just a university student, and called Baryalay on his cell phone. The militants were told by Baryalay that most likely they were looking for him, and that his brother had nothing to do with his job. The militants released Baryalay's brother, warning Baryalay that they would soon find him and punish him for his collaboration with the foreign NGO. Following the incident, Baryalay relocated his wife and daughter to Jalalabad, where they all now live together in the small room he had rented for himself some time before. Baryalay now goes back to his village only on rare occasions.

the multiethnic composition of the Afghan state, every single Pashtun (male) individual is today in a precarious position. Pashtuns have much to lose from a new unfolding political arrangement in the country, heralded by the current Western-backed government. Any change or modification to the state of facts that, until the demise of the Taliban regime, had legitimized the prerogatives of Pashtuns in Afghanistan to the detriment of all other ethnicities, will endanger the maintenance of such prerogatives, and cause the consequent diminution of the intrinsic "value" of the personal identification with the Pashtun ethnic group.[6] This broad sociopolitical dynamics has therefore a repercussion on the psychological dynamics of individual Pashtuns. A diminution of the social and political importance of Pashtuns as an ethnic group in Afghanistan (although often couched in moral terms) may possibly cause a loss of self-esteem and a damaged self-representation in single Pashtun individuals. The mere possibility of this to happen induces fear and anger, interpretable as the result of a perceived narcissistic injury (albeit unconsciously). The bigger the fear of the loss of one's privileged position, the more "obsessive" become the means used to prevent this from happening (i.e., to prevent the narcissistic injury from materializing).

Seen through this lens, the phenomenon seems akin to what psychoanalyst Heinz Kohut defined as "narcissistic rage" (Kohut 1972: 360–400. Kohut mentions explicitly what he calls "shame-prone societies"). The reaction to the specter of a diminished personal value is as uncompromising and violent as it is disproportionate to the threat. Fear becomes omnipresent and pervasive (if unconscious), and any small detail is poised to (almost irrationally) be interpreted as a threat, reinforcing the feeling of uncertainty and fear. Narcissistic rage is publicly enacted through the radical excommunication of any individual who might even remotely be thought to be implicated with those whom people mainly believe to be the sources of the perceived threat (foreigners, and Afghans who work alongside the foreigners, in any capacity).

---

[6] Paradoxically, in discussing this issue some of my Pashtun informants spoke of their group explicitly in terms of "persecution." The perception of the situation was that other ethnic groups in Afghanistan were consciously trying to put down Pashtuns, in order to take revenge and gain control over them. They felt disenfranchised by the unfolding political situation in the country. Such interpretation is particularly interesting because of the dominance that Pashtuns have maintained over the political and social spheres in Afghanistan since the eighteenth century. The fear for the loss of such dominant position (enacted by the rise in power of other ethnic groups) gives birth to a paradoxical sense of victimization. Particularly, Hazaras were seen as viciously intent at attacking Pashtun sociopolitical positions. In this regard, however, a few among my informants managed to express a more "introspective" opinion when they admitted that their fear stemmed from a slightly different realization: "If they get too much power, they will do to us what we did to them."

At the same time, in order to minimize the potential damage, the boundaries of the safe and secure in-group shrink more and more, condemning an increasing number of community members to be ostracized from the in-group and "othered," for seemingly irrelevant missteps (see Campbell and LeVine 1961, LeVine and Campbell 1971, LeVine 2001. More recently, from a political perspective, see Kinder and Kam 2009). In the case of Baryalay, for example, his activities with the foreign agricultural NGO have nothing to do in reality with any political or educational propaganda. His position involves technical assistance to farmers, and the foreign money that the NGO utilizes goes only toward the enhancement of agricultural productivity for local farmers. Nevertheless, the mechanism of "contagion" makes him to the eyes of some community members an "infected" individual, who is by definition influenced by morally corrupt ideas and values. He is thus susceptible to pay the price of the narcissistic rage that erupts in the people, who see him as the incarnation of the threat that they most fear: the loss of social capital and individual self-esteem that derives from Baryalay's status.

As another example, some among my informants had to interrupt any relationship with me midway through my fieldwork after rumors spread in their villages that they were frequenting a foreigner, which was interpreted as a potential proof to their disloyalty to the group, or, worse, of their possible incompatibility with Pashtun morals and political objectives. The fear of violent retribution made them stop meeting me in any way, either in their village or in Jalalabad.

However much Baryalay rejects this kind of mindset and psychological response to a (recent) sociopolitical dynamics unfolding in Afghanistan, he is nevertheless caught by the same phenomenon when it comes to intra-community matters that threaten the supposed moral perfection and Islamic purity of a Pashtun community. Personal misconduct endangered the self-esteem and sense of religious exceptionalism of all those who deeply identified themselves with the community as a whole (see the case of multiple adultery). The response that followed, in terms of narcissistic rage, is similar to that which follows in the previous cases mentioned in relation to the changing sociopolitical Afghan scene. A violent retribution was planned, although it was not yet carried out. These two faces of the same coin seem to be very much peculiar to Pashtun Afghans, at least in the radicalism and uncompromising fashion in which they are enacted. In this sense, Baryalay is still just as much a pacha as he is a Pashtun man. In longing to express his sense of the purity of the in-group, as well as to display his masculinity in protecting it, he demonstrates the depth of his Pashtun enculturation, and the relevance for himself of his "Pashtun self," alongside his pious and peaceful "pacha self." The appropriate norms of comportment for a "true and respectable" Pashtun man still inform and constrain his behavior to a remarkable degree. Not only was his pious Muslim self outraged at what he knew was happening in the villages nearby, but his appropriately Pashtun masculine self saw as the necessary solution to the situation a retaliatory action, which, in line with rawaj in cases of compromised honor, was required to both clear the reputation of the community, and affirm the masculine worth of the men who would be eventually involved in the punitive expedition.

What interestingly emerges from the last passage is also Baryalay's attitude toward women's condition and sexual vicissitudes in a Pashtun environment. It is difficult to gauge at the roots of such interpretation of women's problematic situation, which is by no means a mainstream interpretation among the many Pashtun male individuals I encountered (of all social strata, and from diverse educational backgrounds). To begin with, Baryalay's understanding of his "promise" to his wife not to engage in extramarital relationships is noticeable in and of itself. As he also remarks, in passing, it is very common (upon my experience as well) for married men to have illicit relationships with other women, whether married or unmarried (outside cases of plain prostitution, which are also found). The moral double standard, whereby it is not considered shameful for a man to engage in such relationships, while for a woman it is deemed to bring shame and dishonor upon the whole family, is hardly ever articulated and reflected upon, and usually it is explained away with a simple "that's the way it goes, and how it should be." In one particular occasion, a young unmarried friend of mine from Jalalabad, while discussing with me our own respective sexual lives, reacted with astonishment at the fact that I did not have any sexual relationships with other women since I got into the relationship with my wife. He asked me: "Why did you do that? Are you stupid? It's like eating only Kabuli palaw [a common dish of rice and meat] every day for the rest of your life . . . don't you want to eat something else?" In his opinion, it was clearly a foolish mistake, on my part, not to have taken advantage of the prerogatives that my position as the man in the relationship would have provided me with. He went on by describing a recent sexual experience he had had in Kabul with a young married woman whom he casually met at a job interview. It is well possible that some of the "adventures" described by my informants may have not in fact happened, and I was indeed told by some of my informants that sometimes (and not surprisingly) men make up stories to enhance their masculine reputation among their friends. Nevertheless, the recurrence of these "confessions" is such that, even controlling for possible fake accounts, statistically such phenomenon must be quite common among men. Thus, the attitude that Baryalay has toward marriage, and his wife in particular, is rather remarkable.

One more element worth noticing is what Baryalay says about "natural impulses" toward sexual satisfaction, even for women. He is adamant in acknowledging that women, as much as men, are subject to sexual urges that would be "unnatural" to frustrate. He seems to justify, after all, the behavior of the young wife of the farmer (who in turn has a relationship with a forty-year-old woman). She is young, Baryalay says, and certain impulses are difficult to repress indefinitely. The responsibility for her adulterous actions seems to fall upon the reproachable behavior of her husband, the farmer, who wrongly chose to start a relationship with the forty-year-old woman in the first place. Baryalay is also ready to acknowledge the unfairness of the fact that the liberty he could take for himself in sexual matters, in the case of a protracted absence from home, would never be conceded to his wife. She would be confined to her house under the constant surveillance of Baryalay's male family members. Although he would not allow the situation to develop in any different way, due to the dire social consequences, Baryalay at least seems conscious and aware of the moral double standard that would condemn his wife to a repressed and frustrated sexual life. She lives in a prison, he admits, from which she cannot escape. Such a standpoint is definitely not common among men in Pashtun contexts. Baryalay's empathy and egalitarian attitude toward the opposite sex represents a rarity, as far as my experience goes. I am not able to speculate about the source of such peculiarity in Baryalay's views. Is it the outcome of the penetration of Western-oriented ideas and sociopolitical ideologies into the midst of well-educated, if peripheral, Afghan men? Is it an aspect of Baryalay's pious and inclusive Islamic upbringing as a pacha? Certainly the answer will be overdetermined, and these elements will surely be part of it.

As we have seen above, enmities (dukhmani) and interpersonal conflicts (lanja) are what renders Baryalay's life in the village least bearable. The incompatibility of this aspect of Pashtun life (very common across all Pashtun rural areas, at this point in time in Afghanistan) with Baryalay's personality traits and inclinations was extensively elaborated during our conversations. He described a few incidents that took place in his village to explain more vividly the situation.

I told you, the people in the village fight all the time . . . it's like they enjoy fighting. For example, four or five months ago, the farmer who is my neighbor, and who has a field that borders one of our fields, started cultivating a field that had been left abandoned for many years and borders both our fields. We don't know who owns it, it has been left to itself for a long time. Anyway, in order to reach other fields that we own, we had to pass through this aban-doned field. It's a common thing, there is often a right of passage into the fields of other people [very often cultivated fields are intersected by narrow strips of elevated, non-cultivated ground, in order to give the possibility to others to pass through one's fields freely]. *One day, this neighbor of ours started working the land of the abandoned field, and did not allow us any-more to pass through. We had to go a long way to get to the other fields we owned, beyond the abandoned field. We went several times to this farmer to ask why he was appropriating something that was not his, and to try to make him change his behavior. He did not care, he just said that he wanted to cultivate that field, and that he did not give us the right to pass through. He was arrogant* [mughoror], *and acted like a bully* [zurawaar]. *At that point many other families would have gathered their men and would have ended up fighting with the men of the other family to solve the thing. Our neighbor was disrespecting us* [ihtiram na kawalo], *and our honor* [izzat] *was at stake. But we chose not to fight. We went to some elders in the village, and we called up a Jirga to solve the problem without fighting. The guy ac-cepted the Jirga, and now we are in negotiations. The family of this neighbor is a powerful family . . . this is why he thought that he could do this abuse* [zulm] *without problems. But we have influence too* [nufuz] *. . . people in the office of the district governor know us, they know we are pachaiaan and that we are good Muslims . . . the guy had to accept the Jirga. . . . The elders have already told him that the right of passage is something that he must give to other people, he can't just do as he pleases. . . . We have yet to reach an agreement.*

ANDREA—*So, he accepted the Jirga just because you were also an influential family* [ghat famil way]?

BARYALAY—*Yes, I am sure that if we had been a weak family* [zaif], *he would have not accepted the Jirga. You see, abusive people* [zurawaraan], *if they are powerful, they can do whatever they want, they can even not accept a Jirga. Who is going to force them? The elders do not have the power to imple-ment their decisions* [faisala]. *. . . And even when a powerful person accepts the Jirga, after he has taken away a piece of land that did not belong to him* *from a weak family, a Jirga will only be able to settle things by leaving half of the stolen land to the zurawaar ... the legitimate owner will lose half of his land anyway ... it's not fair ... it's really wrong. ...*

ANDREA—*Does it happen often?*

BARYALAY—*Yes, at least more than it did before. Old people say that it got worse in the past years. ... It's crazy ... one day you go to your field and you might find somebody that is working your land ... they start working a little bit of your land across the boundary. You confront them, and they might say, "Oh, I am sorry, I did not realize," or even, "I don't care, I work this land as mine now." ... What can you do? ... It's bad [da kharab de]. ... There are a lot of zurawaraan now. ...*

The conflict between the different cultural schemata within Baryalay (and the different subjectivities that build upon them) is evident here. He knows the rules of the game in the village, and his subjectivity (his self-image that is premised on pukhto) is shaped by them. Hence, he is aware that letting someone take advantage of his rights and properties jeopardizes the honor of the family (beyond being simply unfair in terms of "common sense"). Such a schema, developed according to the usual Pashtun cultural narrative, resonates strongly with Baryalay, and indexes the relevance of the Pashtun self in him. However, unlike many others in the village, he and his family members (except his father, of course) manage to refuse to conform to the custom that would have self-help (i.e., violence) as the primary resource to tap in cases of compromised honor. The pacha upbringing, and its attending cultural schemata (which have shaped a concomitant subjectivity, self-image), calls for a different resolution of the problem. Crucial for the coherence of his complementary "pacha self," these moral and ethical principles overcome those derived from the Pashtun cultural narrative, and affirm themselves as the main element behind Baryalay behavior in this case. Again, the two selves, the two sets of "meaning-making," that Baryalay is carrying forward are complementary, not mutually exclusive. There is no apparent fragmentation in the overall self-perception (his self-representation) that Baryalay experiences. Where the "pacha self" overrules the "Pashtun self" (or rather, the current, widespread interpretation of the Pashtun cultural norms that Baryalay partly rejects), it is in order to embody a "better" Pashtun, closer to the ideal model of it (or what Baryalay believes it to be). Even more, we have previously seen how the two sets of subjectivities, one based on religious sensitivity, the other on cultural propriety in displaying manliness, managed to interact complementarily in the account he gave of the planned punitive expedition against the adulterous villagers living in his area. Though utilitarian and self-legitimizing as it might seem, Baryalay's management of his multiple selves is positively functional to his psychic balance (notwithstanding the daily suffering and frustrations), and is perceived as providing coherence for that illusion of the unity and univocality of the self that is so necessary to human psychological equilibrium (Modell 2008, Frie 2008a, Bromberg 1996, Stolorow and Atwood 1994).

The legacy of three decades of violent conflict

It would appear that incidents involving abuse and violence, like the one just sketched above, seemingly happen to many on a regular basis. In order to elucidate the chaotic and anxious "everybody-for-himself" village dynamics, Baryalay recounted another incident that had happened a short time prior to one of our conversations. While the three decades of continuous conflict might be at the root of these problematic behaviors, we will see that the last decade in particular has added a peculiar spin to the issue.

> *Some twenty years ago, an old villager who had been left without offspring and wives made a will before dying, in which he donated a big piece of land to the village mosque as* waqf *[a religious endowment]. This piece of land became the graveyard of the village. Six months ago, more or less, the descendants of this man started forbidding people to bury their dead in the graveyard. They claimed that the field was their property, and cut all the mulberry trees that were growing on the lot. They used the trees to make firewood. The villagers became very angry, and I was angry too. It was not fair for them to claim the land as theirs and cut the old trees. We went to the* woliswal *[the district governor] and with him we chose a few elders who would set up a Jirga to discuss the issue. The Jirga decided that the descendants of the old man did not have the right to claim the graveyard for themselves, and that they had to allow people to bury their dead in the field as previously. In the beginning, the descendants of the old man accepted the decision. Then, a few weeks later, they again started to prevent people from using the graveyard. We villagers complained, and waited to decide what to do. One day, unexpectedly, the Taliban arrived in the village. Somebody had alerted them of the problem. I don't know who it was. The Taliban took control of the situation. They had*

their faces covered, I don't think they were people from the area . . . but they were not from Pakistan, they were from here. . . They ordered both parties [the descendants of the old man, and the villagers] *to appoint two religious scholars who would discuss the matter among themselves, and then reach a verdict. Here, in the district, we have some preachers who have gone to school in Pakistan, and got some real degrees from religious schools, so we appointed two of them we knew, and they did the same. Our two mullahs went to one Haqqania madrasa in Pakistan* [a Wahhabi-oriented madrasa]. *The Taliban acknowledged that we found our jirgamaraan* [the Jirga members], *and told us: "Now you have a Jirga. Let the jirgamaraan take a decision, and then re-spect it. It is based on shari'a. If you will not respect it, we will come back and punish you." In the end, the Jirga decided that the family of the old man would have to leave the field to the community, as it was in the past. Nobody went to the woliswal, as we did the first time, or to the police. The Taliban said that anybody who would go for help to the woliswal would be killed. Recently, the family of the old man has requested the opinion of a fifth mullah, to see if the decision taken by the other four was really fair. So far the issue is stopped at that. The Taliban have not come back yet, but I am sure that somebody will call them up again. You see how it goes today, the mullahs have much more importance than in the past. Religion has become more important than rawaj* [custom] *in many cases. When I was a kid, I remember jirgas were made up of only elders . . . the mullah came later, only to make a ritual at the end of the Jirga, so that everybody could say that the decision had been taken according to shari'a. Now the Taliban make you have a mullah as a jirgamaar. But their ideas are wrong, this is not Islam, what they are imposing . . . it's politics. . . . They want to look like they are respecting rawaj and the Jirga, but in fact they are forcing their ideas of religion on the people.*

While Baryalay participates in the indignation of the villagers at the usurpation of the waqf communal property by the family of the previous owner, for the sake of their own personal interests, the development of the facts still leaves him extremely frustrated, for several reasons. The abuse that the family of the old man perpetrated against a religious property for communal use is one more clue for Baryalay to the unbearable situation of everyday life in the village, as well as to the deterioration of the moral fabric of the average Pashtun man. He sees that his ideal model of the righteous man who follows pukhto, embodied by the khan who pays the quotas of other villagers in order to build the bridge for the enjoyment of the whole village, is drifting away from the current social landscape. Likewise, the more recent "religious" bend that social relationships have taken does not leave him comfortable either. It represents a religious fervor that is dependent on, and subservient to, the influx of Islamic radicalism brewing in Pakistan since the late 1970s. It dramatically migrated to Afghanistan during the anti-Soviet war and, especially, after the various waves of returnees came back from the refugee camps in Pakistan after the Soviets' retreat. The Taliban, as Baryalay knows too well, were but one product of such dynamics, and today's "Neo-Taliban" (see Giustozzi 2008) are the updated version of the same phenomenon. The insurgents who now carry out military operations against the Afghan government and the foreign troops are also mostly imbued with (among other things) a religious fundamentalism that reminds Baryalay of the radicalism springing from the Pakistani Wahhabi madrasas, once promoted by the Taliban state apparatus of the late 1990s. It does not certainly represent the kind of Islam that Baryalay was educated into as a pacha. Thus, this incident indexes two main thorns that Baryalay feels in his side, and that poison his life in the village: the unacceptable way of expressing one's masculinity, based on violence and abuse, and the "distorted" interpretation of Islam that is gaining the upper hand in the minds of his fellow villagers.

The increased frequency of violent antigovernment activities by groups that loosely self-identify as "Taliban," or that operate without affiliation with broader-based insurgency movements, is reviving a situation of widespread social insecurity and violent deviant behaviors that many people associate with the civil war of the mid-1990s. Ruthlessness and *zulm* (cruelty) are coming back to being a part of the daily moral landscape in both urban and (especially) rural areas. Those responsible for such violence seem to behave in accordance with a new "state of exception" (in Giorgio Agamben's terms [Agamben 2005]) justified by the occupation of the country by thousands of foreign troops. A novel "holy war" authorizes a new suspension of the moral and ethical guarantees that ideally would belong only to a "normal" historical contingency. The same process that in the 1980s started the shift in behavioral patterns, and authenticated new standards for moral values, which still hold valid in today's Pashtun masculine environment (Fredrik Barth's feedback effect [see Barth 1966]), is developing in the current situation of fierce opposition to the ongoing multinational military effort in Afghanistan. Expectations for the expression of manliness and valor are once again being renegotiated, exacerbating the already harshened standards shaped by the three decades of previous conflicts.

The rural district where Baryalay lives, Pachir-wa-Agam, is one of the epicenters of such process. Removed from the influence of the administrative provincial center, Jalalabad, and far from the main arteries of communication, the district has never experienced a stable and firm government control since 2001. Between 2010 and 2011 Pachir-wa-Agam experienced a severe deterioration in its general security situation that included a steady encroachment on its territory by elements that pursued a violent opposition to the Afghan government and the multinational military contingent. Baryalay often started our sessions with a chronicle of his trip from or to his village by car. He would give me details of the events that had taken place during the latest trip. The picture he portrayed of the general situation in the rural areas where he lived and drove through was bleak. Police stayed holed up in their compounds along the main road from Jalalabad to Pachir-wa-Agam, appearing at the site of an incident only after the fact, only after all the insurgents had fled the scene. Illegal checkpoints by antigovernment forces were growing more frequent, and more ruthless. The Afghan army had no presence in the whole area. Between 4 p.m. and 8 a.m. all roads, and the countryside as a whole, were in total control of the insurgents. He often repeated a saying that apparently was becoming rather popular among locals: *pa wraz ke, de hukumat dawlat de, pa shpe ke, de Talebanu dawlat de* (during the day it's the government's country, during the night it's the Taliban's country). And this still seemed to me an understatement, given that reports of insurgents' activity in broad daylight were pretty common (as I was unfortunately able to confirm firsthand more than once). Indeed, I witnessed the multiplication of episodes that demonstrated the heightened level of violence and how the moral yardstick was changing, against which values like social respectability and qualities such as manliness became measured. Baryalay, like many among his peers, was heavily affected by the perceived change in the general atmosphere. In one of our sessions, he recounted one of the most troubling among these incidents.

Last week, while in a taxi [often a private car that carries people for money] *going back to my village from Jalalabad, people were talking about an incident that had taken place on the same road just a few days earlier. They were telling how a fuel tanker was attacked by the Taliban on the way to Pachir-wa-Agam, near the border with Chaprihar* [the neighboring district]. *They hit the tanker with an RPG-7* [a rocket propelled grenade], *and it went up in flames. The driver jumped down from the vehicle to save himself. The Taliban caught the driver right there on the road. They told him to follow them, but he refused. He said that he would rather die there. So they put him on his knees and shot him in the back of the head with a rifle. Then they left. The other day, while coming to Jalalabad, we stopped to get something to drink and eat at the place where this incident happened. There were food stalls and a small restaurant. I spoke with one of the owners of the stalls, who witnessed the fact, and asked him what had happened. He confirmed the story, and told me that the truck driver looked like a young man, little more than a boy. He could have been seventeen or eighteen, he said. They shot him on the road, in front of everybody, without mercy [rahmat], without shame [sharm]. The owner of the stall was angry at the people who did this. He said it was not humane to do so [insanyat na de], that this was against Islam. A lot of people feel the same way. They are upset at this incident, and many others. People who do these things [i.e., the insurgents] are not human, and are not Muslim. They are crazy. I hate them [zma bad ye razi]. If I was a little more powerful, and if I had friends who would follow me, I would find these people and kill them. I would punish them for what they have done to that boy.*

ANDREA—*But if so many people are against the actions of the Taliban, why don't they do something against them?*

BARYALAY—*In the past, some villages tried to rebel, to do something against these people. But they have been punished, the insurgents [yaghyaan] have gone to those villages and killed the people that had organized action against them [arbakay, or groups of self-help]. People now are scared, and do not dare to do anything anymore.*

ANDREA—*Are these insurgents [yaghyaan] local people, or are they from somewhere else?*

BARYALAY—*No, usually they are from the area . . . we can tell from the way they speak . . . we can't see their faces when they stop our cars, or make checkpoints, because they cover them with a scarf [zaader], but we can hear the way they speak . . . they are mostly from these areas. There might be someone from outside sometimes, maybe people who bring them money, supplies, weapons, but those who do things on the ground, they mostly are from here. You know, they are normal people. . . . I mean, they have normal jobs in the bazaar, or in the fields, or are jobless. When the order comes, they leave their occupation, carry out the attack, and then come back to whatever they do for a living. That's why they cover their faces when they operate . . . they know that there are people that could recognize them, and it could be dangerous for them . . . you know, maybe someone reports them secretly to the Afghan army, or to the foreigners, and then there is a drone attack, or an army attack, or whatever . . . they keep their identity secret [pat]. . . . There is a lot of zulm [cruelty, abuse] these days, much more than in the past, as far as I can remember . . . it's a horrible situation, you hear of things happening almost every day . . . a few weeks ago, on the border between Pachir-wa-Agam and Chaprihar the body of a man was found beheaded. It had been half eaten by wild animals, those who killed him cared to bury him only in part . . . it was found by some people walking down the road. . . .*

Baryalay paused for about half a minute after recounting this story, and I did not interrupt his thoughts. Then he addressed me again with a different topic.

BARYALAY—*You know, one time, when I was in university in Khost. . . . I used to go often back and forth to Peshawar to see my brother who lived there . . . one time I was going back to Khost from Peshawar, and in my taxi there was a friend of mine, with whom I was making the trip back to university, and a man with his sister. He must have been around thirty years old, and his sister I could not see, because she was covered by her burka. They were* kuchyaan [nomadic Pashtuns], *they were very poor. I was sitting in the back seat of the car, beside the man, who was between me and his sister, and my friend was in the front seat. The woman was clearly sick, she was coughing from time to time, and she was mumbling something to his brother's ear often. At times she would stop coughing or talking to his brother, at which point he would shake her, as if he was trying to wake her up or ensure that she was still alive. We had to change cars a couple of times during the trip, because there was no car that went from Peshawar straight to Khost. At the first change, the woman was sitting on the side of the road, with her head down, without moving, and his brother was beside her. I asked him if there was something wrong, and if they needed help. He said that his sister had tuberculosis, and that they had gone to Peshawar to see a doctor, but it was too complicated and expensive, so they decided to go back to Khost and have her see a local doctor. I felt very bad for them, and I talked to my friend. We offered them some money to help them with the medical expenses, and told them that they should go back to Peshawar. The man refused, and did not take our money. We continued the trip, but the woman was getting worse. She moved and talked less and less. During another change of car, finally she stopped moving, and her brother told us that she was dead. They remained there, and my friend convinced me to keep on with our trip, because he was scared of the place we were in . . . it was Waziristan, and he was afraid of the people there. He said they were savages, dangerous, and they should not be trusted. So we drove back to Khost that day. . . . I felt bad afterwards, very bad. . . . I became ill, and had to go to Peshawar to a hospital. . . . I stayed about a month in the hospital, and then I got better. . . .*

ANDREA—*What happened to you? Why were you sick?*

BARYALAY—*I don't know, I felt very bad, I could not sleep, I could not eat. . . . I don't know what happened to me. . . .*

ANDREA—*Did you feel responsible* [masul] *for what had happened to the woman?*

BARYALAY—*I don't know, maybe, but I felt that it was wrong that she died . . . she shouldn't have died. . . .*

ANDREA—*Why are you telling me this story now?*

BARYALAY—*I don't know, it just came up to my mind. . . .*

I interpreted the association that Baryalay spontaneously made between what he had just told me about the security situation in his district (i.e., the attack on the gas tanker, and the beheaded man), and the story about the death of this woman on the way to Khost, as Baryalay's way of unconsciously communicating to me the sense of deep discomfort, claustrophobic anxiety, and moral disgust that he perceived about his social life in the village, as well as his personal condition of being virtually trapped in that state. We did not discuss the deep roots of his otherwise inexplicable, depressive reaction after the disturbing death of the woman during his trip to Khost (i.e., his hospitalization). Rather, my attention was directed to the meaning of the associative link that he had expressed by spontaneously coupling the story of the killing of the truck driver with the death of the woman. The point was not in the reasons for his hospitalization *at that time*, but in the reasons for his thought association *during the session with me*. The association itself had a meta-meaning that went beyond the motives for what had happened to him during and after the trip to Khost. Baryalay was trying to tell me something about himself in the here-and-now through linking together two episodes in his life that apparently had little to do with each other. What these episodes had in common was that their juxtaposition was used by Baryalay to convey information to me about his state of mind and emotions in the present. I could not help but feel the sense of helplessness and impotence that

Baryalay must have felt overwhelmed by, every day, because of the kind of life he was forced to conduct in his native village. In his associative thought process, being crammed in a small car with strangers, without the possibility to move or "escape," having to witness the agony of the last hours of a woman's life, without the possibility of doing anything to help her, running the risk of remaining stranded in a strange place among "dangerous" people, were all images that I believe Baryalay unconsciously used as metaphors for his own suffering and despair in the village (and in Afghanistan as a whole). The latter were epitomized by the dramatic description of the public execution of the young truck driver. It conveyed to me the oppressive state of mind of a villager like Baryalay who is every day surrounded by possible enemies who hide their faces while inflicting violence, only to come back to their homes, perhaps as close as the compound next door. The juxtaposition, or association, of the two stories speaks to the profound feelings of hopelessness that Baryalay confronts as a moral being (culturally expressed in terms of his religious pacha upbringing), as well as a masculine being (culturally expressed in terms of his "Pashtun-ness"). Within a quickly changing landscape of religious morality and performative masculinity, Baryalay finds no easy avenues to publicly express himself simultaneously in a moral and masculine way. Indeed the seeming impossibility to find such an avenue leaves him in a state of painful impotence, as conveyed to me in his unconscious associations during our conversations.[7]

Perceiving oneself: "authenticity" and intimacy

Baryalay's sense of powerlessness (and, likely, masculine inadequacy) is no doubt increased by the fact that also his wife is apparently unhappy with her life in the village. She would like to leave just as much as he does. Baryalay cannot do much to mitigate his wife's discomfort. The two had an arranged marriage, as is customary, some three years prior to the start of our sessions, and have a daughter who at the time was about two years old. Baryalay's wife is an illiterate young woman of twenty-three, also from a pacha background like himself (pachaiaan tend to be an endogamous group), yet from

---

[7] Elsewhere, I elaborate on the "hauntology" (Good 2015, Derrida 1994) of the experience that Baryalay had on his trip, and on the dissociative features that the psychic processes displayed by Baryalay in this exchange with me possibly entail (Chiovenda A. in press).

a different district in the province. The two families are not related by kinship, but rather came to know each other through the close network of acquaintances that all Pashtun pachaiaan maintain throughout Afghanistan. While Baryalay always acknowledged the many downsides of arranged marriages, he consistently declared that he was lucky, because his marriage was proceeding satisfactorily and he and his wife did not have problems or frictions. In his opinion, they adjusted well to each other and cared sincerely for each other, which apparently does not happen very often. Both kept as a priority the good functioning of the marital relationship, Baryalay explained, and he considered himself and his wife a happy couple. On my part, I can say that during our sessions I had the opportunity to witness countless phone calls between Baryalay and his wife, initiated by either one. In fact, in comparison with the frequency with which I witnessed intra-couple phone calls among other informants in similar circumstances, I can certainly consider that Baryalay kept himself in contact with his wife much more often than average. Often the calls represented just a quick check on how the other was doing, and, especially on Baryalay's wife's part, on whether he had had any problems on the road to Jalalabad, which was to her—and rightly so—a very concerning matter. At other times the conversations would center on an update on issues that the household might be going through, fresh news from the village life, or simply the need to keep in touch with each other. The conversations' tone, especially on the part of Baryalay, was always very soft and respectful, lacking a domineering mode that I had heard often in other informants. In one of our sessions he said:

My wife and I have been lucky, because we like each other, we get along together well . . . many others are not so lucky. . . . Relations between our families are good, she can go back to her parents' house any time she wants [for visiting, that is, which is sometimes not a given, and mostly depends on the husband's attitudes toward his wife and her family]. *We take decisions together in the house, I listen to what she says, and she listens to me, and then we take a decision together.*

ANDREA—*Do you think she is ok in your family's village?* [as is usual in Pashtun society, also pacha families are patrilocal].

BARYALAY—*Well, I think she is ok, she adjusts well. . . . I mean, we are pachaiaan, and she is as well, and she understands that pachaiaan do things a little bit differently from other Pashtuns, I explained this to you . . . she understands the importance of our traditions.*

ANDREA—*What do you mean?*

BARYALAY—*Well, you see, for pachaiaan it is customary* [rawaji] *to keep a stricter purdah* [female segregation] *than other Pashtuns . . . pacha women do not run errands in the village, they stay in the house, go out less often than other women do . . . and when they go out, they are never alone and wear a burqa also inside the village, while other women moving inside the village usually wear only a hejab* [a head scarf].

ANDREA—*Does she ever talk to you about her feelings, what she feels about living in the village?*

BARYALAY—*Yes, sometimes, when we are alone, she complains a little about our life. She says that she is tired of all the risks, the problems, that she would like to leave Afghanistan, go somewhere she could be more free* [khoshi]. . . .

ANDREA—*Did she ever go anywhere else, like Peshawar, or Kabul?*

BARYALAY—*No, she only lived in her village, and my village.*

ANDREA—*So, how does she know that other places are more "free," that she would feel better living somewhere else?*

BARYALAY—*Well, several members of her family left Afghanistan a long time ago, and now they live in Australia, and I think also Canada. They call her on the phone from time to time, and every once in a while they come back to Afghanistan to visit. So she talks to them a lot, they tell her about their lives, what they do, their jobs . . . she listens as if she was listening to a tale* [nukkal]. *She also tells them about her own life, they discuss about it, I guess that she realizes the difference . . . so she dreams of going away from here.* . . .

ANDREA—*So your wife is not exactly happy.* . . .

BARYALAY—*Well, she adjusts well to our situation, but she would like to live differently, I guess.* . . .

ANDREA—*How does this make you feel? How do you feel about your wife being unhappy* [khapa] *in Afghanistan?*

BARYALAY—*I feel bad for her. . . . I understand why she does not like it here, I too wish we could live somewhere else. . . . I wish I could do something for her . . . but what could I do? It's not my fault if we cannot leave, right? I have a job, but that's not enough to give us the money to leave . . . and then I have responsibilities with my family.* . . .

ANDREA—*But imagine you could leave, go to Canada, imagine, right?* [fars kra, ka na?] *You know that in the West certain rules of behavior* [barkhord] *are different than in Afghanistan, no? For instance, women can leave the house as they please, and they often can find themselves a job. What would you do if your wife told you that she wanted to live like that?*

BARYALAY—*No, I know that in the West things are different. . . . I have no problems with my wife going out of the house, and even finding a job . . . the issue would be with my family back in Afghanistan. If they knew that my wife was working out of the house, among strange men, without me to check on her, they would make problems for us, they would give us* peghor [the act of vociferously complaining with someone for contravening moral norms] . . . *that would make me and my family dishonorable* [beghairata] . . .

The relationship that Baryalay has with his wife is certainly the most intimate and tender among all the informants that I worked with in Afghanistan. During the many hours I spent with other friends, very rarely have I witnessed phone conversations between husband and wife, and when I did, the general tone was always less "equal" (for lack of a better term) than the one I found in Baryalay. That they "take decisions together," as Baryalay stated, was more than rhetoric geared toward the Western listener, whose mindset and expectations are widely known and anticipated among Afghans, especially in the delicate realm of gender relations. I actually heard Baryalay more than once discuss with his wife by phone the best course of action to take about various issues—from what to do when their daughter fell sick, to how to organize a certain family celebration, or whether he should cut short his stay in Jalalabad to go back home to the village and give help to his wife with the household management. It is such intimacy, and emotional closeness, that makes me believe that Baryalay suffers the frustration of not being able to fulfill some of his wife's wishes, which he deems legitimate and understandable—for example, longing for a different place where to spend their lives. Against the background of a culturally shaped masculinity requiring that he be the pulling force within the nuclear family, it is precisely the attention that Baryalay seems to sincerely devote to his wife's well-being that renders his incapability to "deliver" most hurting and humiliating. If, on the one hand, he feels the responsibility for being the "provider" (in this case, of a better life for his wife), on the other he feels trapped by his more traditional role as a productive member of his original paternal family. His father has acrimoniously estranged the rest of his family from his purview, and two of Baryalay's brothers (the shopkeeper in the village, and the younger university student in Jalalabad) need financial support. Adding to this, Baryalay cannot help but imagine his family members' gaze following him to the place where he would eventually move if he had the chance to do so. One of the main concerns in any Pashtun man's life (that of social censorship and public reprobation) haunts Baryalay even before anything has the chance to happen—and has the power to cripple blooming desires and future plans. What would they think if they knew?, he imagines of his family members and his fellow villagers at the news that Baryalay's wife now works in an office in the West, surrounded by strange men, without even the "supervision" of her husband. During the conversation in which Baryalay expressed these doubts, fears, and hesitations about the future, his facial expressions and his tone of voice made me feel a sense of helplessness, and almost desperation. This emotional countertransferential reaction on my part was, I believe, an empathic and attuned consequence of the evident, intense frustration that was exuding from Baryalay's personal account.

It is somewhat ironic that Pashtun men's life is in fact lived in a sort of social panopticon, where everyone is at the mercy of everyone's else's gaze (as well as unconscious feelings and projective identifications), and yet (or, rather, because of this) at the same time each person's private life is replete with interstices that they strive to keep hidden (and hence safe) from anyone's knowledge. Such interstices are both the source of continuous frustration and dissatisfaction (for they no doubt contain conscious and unconscious unfulfilled wishes), but also the locus for the "detection," and the discovery of personal agency and autonomous emotionality. In the case of Baryalay, for example, his longing for a radically different change in lifestyle, and even in physical place of settlement, is kept secret from anybody in his community. If he told anyone about his actual dreams and wishes for his and his wife's future, he likely would get himself into trouble.

> You cannot say anything anymore to anybody in the village. You don't know whom you are talking to any longer. . . . It could be someone who does not like you and is connected to the Taliban. . . . You say the wrong thing, like "I think this is not the right way of doing Islam," or "I wish I could live somewhere else," and you risk to be called a bad Muslim, or a spy, or a Christian . . . you cannot trust anybody, there are no friends anymore. It only takes a wrong word to the wrong person. . . .

Such a condition of constant uncertainty and insecurity has deleterious effects on everyone's psychological equilibrium, as Baryalay underscored several times with me. He is apparently balanced enough, and strongly enough enculturated into his "quietist" pacha religious background, to evade the worst consequences of this continuous state of tension and uncertainty. Others are not, and there lie the roots, in Baryalay's opinion, of many negative reactive behaviors that he highlighted in our conversations, such as interpersonal aggressiveness and domestic violence. On the other hand, though, precisely the persistent inability to publicly express and pragmatically pursue certain private strivings and wishes puts Baryalay in clear contact with such private longings. Precisely because so vehemently curtailed and hindered by a strict and self-policing social environment, such private wishes emerge and surface to Baryalay's awareness and consciousness (as shown by his bitter resentment, sense of helplessness, and longing for a different fate).

This phenomenon, I believe, may represent what psychoanalyst Wilfred Bion has termed "truthfulness" to oneself (Bion 1967a. For a discussion of Bion's interpretation of what the patient [or informant] might experience as "true," see Ogden, 2004: 293–298), and what I would rather call "authenticity." In this sense, authenticity is not to be intended in its cultural connotation (for the many aspects of which, see Lindholm 2008), but rather as the outcome of a positive, functional psychological dynamic, which brings the individual to a clearer acknowledgment of his or her deeper feelings and desires in relation to a specific life contingency (i.e., the here-and-now). Knowledge starts with sensing, Bion argues, with an emotional elaboration that remains "raw," unprocessed, until unconsciously "cooked" (to use an apt term by Italian analyst Antonino Ferro. See Ferro 2005, 2006a, 2006b) by the mind, and stored in an unconscious form, only to tapped into when necessary to the individual's psychic equilibrium. The capacity of each individual to "recall" stored emotional elaborations (which is not a given by any means) is what renders life experiences meaningful and functional to one's inner dynamics. In this sense, being able to "feel," to perceive one's own deepest wishes and desires, in the face of environmental adverse cues, means to be truthful to oneself, to be "authentic" to oneself *in the here-and-now*. Thus authenticity, from this standpoint, is not an ontological, absolute category, but rather an experiential and contingent subjective state. Not always, and not everybody at any given time, in Bion's view, manages to complete such a process of personal growth (Bion 1962, 1967a). Failing to do so leads to inner suffering and deleterious psychic ramifications. Upon my conversations with Baryalay, and following Bion's suggestive proposition, it seems to me that he was able, to a certain degree, to tap from the "cooked," processed emotional material he had stored in his unconscious, to come into contact with the deeply frustrating feelings about his life conditions, and his conflictive desires for the future. In these terms, Baryalay was showing to be "authentic" to himself at that very conjuncture in his lifetime.

Conclusion

From the detailed analysis of my conversations with Baryalay, several issues, I believe, emerge, which had been already adumbrated by the discussion of Umar's and Rohullah's life histories.

First is the fast-paced metamorphosis of cultural idioms of masculinity among Afghan Pashtuns in the area where I worked. Baryalay quite clearly expressed the view, advanced also by Umar and Rohullah, that the three decades of war have taken a harsh toll on the way Pashtun men interpret, live, and perform their sense of masculinity. Cultural idioms, within this purview, are, and have been for a long time, shifting away from pre-conflict standards. The long period of relative peace that followed the demise of Amanullah Khan from the throne in 1929 was abruptly interrupted by the Communist Saur Revolution first, in 1978, and later by the Soviet occupation of the country (1979–1989), which culminated in the civil war (1992–1996) and creation of the Taliban regime (1996–2001). All the informants whose life histories I have so far analyzed perceived that the moral and ethical standards on which pukhto (the Pashtun customs) had been premised for the past three generations were impacted strongly and perhaps irreversibly by the traumas and the high degree of violence produced by these continuous conflicts. The idea of nartob (manliness), which is so much a crucial aspect of pukhto for any Pashtun male individual, was in my informants' opinion accordingly affected. I have underlined the feedback process through which, in my opinion, aberrant, violent behaviors were justified by a "state of exception" instituted by the conflict environment, and then became routinized and institutionalized, giving rise to a new set of expected moral standards and models for personal and collective action—at least in the eyes of those Pashtuns born and raised in the midst of those wars.[8] I have tried to highlight the psychological dynamics to which single individuals are subjected when confronted with such state of facts. The individuals I chose to present so far are men who confront a reality that they do not completely accept, and which they have lived, or still live, at least partially in opposition to their deepest wishes for

---

[8] "Morally legitimate" is preceded, chronologically and hence historically, by "pragmatically necessary." It is precisely the subtle and imperceptible passage from the second to the first condition that I am arguing happened within Pashtun society over the decades of conflict. It is a corollary to Pierre Bourdieu's articulation of practice theory that I believe Fredrik Barth has *ante litteram* acutely perceived (Barth 1966). In Charles Lindholm's words: "Men may well know that rule by pure force is wrong, but must proceed according to the demands of force or else submit to others who are less squeamish" (Lindholm, personal communication, October 2014).

the future and self-representation in the present. The shift in meaning that idioms of masculinity have undergone since the beginning of the "state of exception" (1978) are taking a heavy toll on the inner processes and subjective states of men like my informants.

To such dynamics is tied a second issue that I think emerges from the analysis of the material I presented. The private conflicts my informants verbally articulated during their conversations with me revealed interesting ways in which they managed the contrasting inputs they received from cultural cues and social constraints, and "private" wishes, emotions, and strivings. I have interpreted the accounts they gave me of such conflicts as showing the presence of multiple layers of subjectivity, or selves, which were coexisting and complementary, although at times also in competition with each other. Each layer of subjectivity, or self, is constructed by the individual, I argue, to respond to contingent real-life conditions, on the basis of one's own enculturation process (i.e., cultural schemata), as well as one's idiosyncratic elaboration of, and reaction to, environmental cues.

Using a concept borrowed from psychoanalysts Joseph Sandler and Bernard Rosenblatt (1962), I would argue that the representational world of each of my informants remains surprisingly coherent, and is perceived by them as imbued with characteristics that support the necessary "illusion" of a unity of the self (in spite of the many episodes of disorientation, anguish, and internal conflict that they suffered, and which they were able to convey to me in conscious or unconscious manners). This seems to happen because, as suggested by Sandler and Rosenblatt, it is the "self-images" which each individual construes of himself that change and shift in order to respond to the stresses and challenges of lived life. The underlying, broader self-representation of oneself, on the contrary, takes advantage of these shifts among self-images that produce psychological balance, and for this reason manages to be perceived by the individual as coherent and "unified." The shifting self-images, in turn, may be interpreted, following psychoanalyst Antonino Ferro's suggestions, as the result of the unconscious tapping into "cooked" experiential baggage (i.e., the transformation of raw, inchoate, sensual, emotional experience, into an accessible pool of knowledge. Ferro 2005, 2006a, 2006b), which serves for the individual the purpose of facing and overcoming the pragmatic, embodied contingencies of daily life (in an ideal, "healthy" scenario, of course).

Furthermore, as Robert LeVine has suggested (LeVine, personal communication, May 2014), a representational world is not peculiar exclusively to single individuals. Culture as well consists of representations. Different cultures produce different representational worlds. We might define these cultural representational worlds as composed by "idioms," "schemata," or "narratives" (as I have so far done). Moral standards and ethical injunctions, that among Pashtuns define "how to be good at being a man" (in Michael Herzfeld's words. Herzfeld 1985: 16), are part of the Pashtun cultural representational world. By the same token, the interpretation and negotiation of, reaction to, strategizing behaviors upon, and adjustments to, this portion of the Pashtun cultural representational world on the part of my informants define their own shifting personal self-images (subjectivities, or selves), as well as their own broader, coherent self-representation as a whole (i.e., the coherent coexistence of their self-images). It is easy, at this point, to see how the cultural and the personal representational worlds are not only constantly interacting with each other, but also partially overlapping (albeit to different, idiosyncratic degrees). In fact, cultural representations become motivating and meaningful precisely because they have the power to create emotional responses in individuals, while in turn individuals necessarily formulate their own representational world and self-images in cultural terms (though, again, idiosyncratically. See also Hollan 2000).[9]

A final issue that emerges from the analysis I have developed so far concerns the idea of personal agency and autonomy. Few anthropologists and cross-cultural psychoanalysts in the past have repeatedly underscored the fact that, contrary to what many others had suggested, it is just within those sociocultural environments where strict rules and norms of comportment are in place, and where social self-policing and public control are strongest, that individuals manage to "detect" more clearly their private wishes and strivings, and somehow act upon them (Doi 1981, 1986, Lindholm 1997, Roland 1988). I would elaborate further by saying that against the background of strict intra-community social surveillance, individuals appear most able to remain "in contact" with their inner subjective states (or those self-images that at that moment in time are most relevant for the coherence of their representational world as a whole). More recently, Bambi Chapin has presented ethnographic material arguing for the same principles (Chapin 2013: 152–158).

---

[9] I wish to thank Professor Robert LeVine for suggesting and discussing with me these ideas, as well as the article by Sandler and Rosenblatt that I mentioned.

The three informants that I presented so far showed their idiosyncratic strategies (whether conscious or unconscious) for coping with their alternative and complementary (and at times conflictive) subjectivities. To varying degrees, they expressed in conversation with me (and seemed to have perceived in the past, upon the retrospective recollection of their experiences), an extent of personal agency and autonomous self-awareness that would have escaped detection by the outsider observer faced simply with their outward behavior. Particularly, Rohullah seems to have managed to remain "in contact" with the "authenticity" of his self, that is, with the overall meaning of his self-representation as a whole (not necessarily unchanging) at particular stages in his life, in spite of the shifting self-images that he construed for himself in order to adjust to the progression of his life contingencies, and in order to maintain a "healthy" psychological balance in the face of existential hardships. We have seen, on the other hand, how the representational world of Baryalay seems to have been constantly characterized by the preeminence of cultural idioms derived from his enculturation into the pacha religious environment. Whereas this seems to have been a *fil rouge* that held together his overall self-representation, he shows to be nonetheless strongly shaped by the alternative and shifting self-images that different cultural cues have imprinted onto him, such as the norms and injunctions on how to live according to Pashtun's nartob (manliness), and the crucial values related to ideas of honor (izzat, ghairat), and shame (sharm).

Such dynamics in Rohullah and Baryalay have highlighted how both individuals gained (or, better, won for themselves) a heightened awareness of alternative avenues for behavior and being that they felt more syntonic with their self-representation, even though (or rather, because) they were not allowed to endorse those behaviors publicly, due to social and cultural constraints. In other words, I am arguing that precisely the impossibility of publicly expressing and displaying certain preferences, wishes, and behaviors sharpened my informants' awareness of the underlying shape and form of their current representational world, rendering them able to, if often subversively or covertly, maintain and give voice to a personal agency and autonomous being that would seem at first glance unlikely, given the sociocultural environment in which they are living. It is this capacity, on the part of my informants, that I call "authenticity."

5

Rahmat

The Dilemmas of a "Perfect" Pashtun

Prologue

Sama Khel is the rural village where I have worked most often during my fieldwork in Afghanistan. It is located in an area of Nangarhar province that is mostly inhabited by Shinwari Pashtuns, a particular sub-group of the broader Pashtun ethnic group. Shinwaris, as it often happens in segmentary lineage systems, are in theory all related through an apical ancestor, in turn related horizontally to the other apical ancestors of different Pashtun sub-groups. In reality, as has been shown in ethnographic literature (see, e.g., Kuper 1982, Hammoudi 1996, Barth 1969), the boundaries of ethnic entities tend to be more permeable than their members portray them to be in ideal terms. Although almost all those who inhabit the area claim to members of the Shinwari sub-group, the membership is often contested and subject to disputes.

Sama Khel lies some twenty kilometers west of the Afghan-Pakistani border, fifty kilometers east of Jalalabad, and about five kilometers south of the main throughway that connects Peshawar, Pakistan, to Jalalabad. As a result of historical circumstances (among which the most relevant was the creation of the Durand Line in 1893), the Shinwaris now live on both sides of the border in Pakistan and Afghanistan. Afghan Shinwari families often have relatives and private dwellings in Pakistan, and regularly visit Peshawar when in need of serious medical assistance, or specific goods that cannot be found in Afghanistan. I discovered with surprise that it is extremely common for a rural villager of the Sama Khel area to have been several times to Peshawar (or even to have lived there for a period of time), but to have never visited Kabul or any other major city in Afghanistan.

Baked in the spring and summer by a scorching sun, but blessed in the (late) fall and winter by a pleasantly mild climate, Sama Khel and its surroundings take advantage of the dense network of artificial canals (*karez*)

Crafting Masculine Selves. Andrea Chiovenda, Oxford University Press (2020). © Oxford University Press.
DOI: 10.1093/oso/9780190073558.001.0001

that the Soviet-funded government built in the late 1970s and 1980s to channel the waters of the close-by Kabul river, as well as those flowing down from the nearby Spin Ghar mountain range. Irrigation makes the fields adjoining the canals lush with orchards and various field crops. Away from the canals, however, the land is barren and dry as far as the eye can see. Precipitation is extremely low all year round, and the intense heat quickly evaporates the little rainwater that leaches through the first layer of soil. By way of comparison, Jalalabad receives ca. 7 inches of rainfall per year, while Phoenix, Arizona, receives ca. 8 inches, and Riyadh, Saudi Arabia, almost 4 inches. Sama Khel is slightly drier and hotter than Jalalabad.

People living in Sama Khel, and its district capital, Angur Bagh, are primarily employed in farming (whether as landless individuals laboring other people's lands, or as landowners, reaping the fruits of the fields they do not personally farm), in commercial activities in the district's bazaar, in the local public administration, or they are jobless (of whom, at the time of my fieldwork, there was a considerable number). Pastoralism is done on a small scale, and often for reasons of subsistence only. The average extension of private land that an individual owns is not large, usually a few jiribs, although there are some families who own, collectively, a fair amount of land. Ascertaining precisely the extent of land holdings my main informants possessed was nearly impossible, given that all of them proved extremely reluctant to discuss with me the real facts about their wealth (or lack thereof), and, more important, because the fields they possessed were in most cases scattered around the countryside surrounding the village, where the little security that the village offered vanished completely. I was therefore never allowed to accompany them to their fields—it would have been dangerous (or, at least, not advisable) for them to be seen surveying the territory in the company of a stranger, whom some in Angur Bagh even knew was a foreigner.

Sama Khel is a village-fortress, typical of Pashtun areas (see Szabo and Barfield 1991). Its population, about 450 people, is composed entirely of people from the Waraki *khel* (lineage). They are all related by blood to various degrees—the exception being the occasional brides from different lineages in the district, who come to live in the village with their husbands' families. Different family houses cluster together, giving rise to an uninterrupted outer wall. On the inside there are open courtyards, alleys, separation walls for each house, communal wells, and other features proper to a common village structure. All is insulated from the outside world, however. Since the segregation of the sexes is strictly enforced for those who are not closely related by blood, and hence allowed in theory to marry each other, I was never allowed to visit more than the male guests' quarters (*hujra*) of each house, and the courtyard around them. I have never seen any of the women living in the village of Sama Khel.

Most of my time I spent in the main, large seventy-year-old village hujra, where all the major male figures congregated in the morning, and after working hours in the afternoon. I also slept in the same main hujra when I stayed overnight in the village, together with some other occasional guests, and a few local villagers who wished to keep me company until bedtime. My mobility around the area was limited by the overall security situation of the surroundings. My hosts could only accept to take so much risk for the sake of my academic research. The performance of the proverbially uncon-ditional Pashtun hospitality—*melmastia*—is in fact much more conditional than is often portrayed in past ethnographic accounts.[1] Because people wan-dering alongside me through Angur Bagh's bazaar or even the dusty roads around the villages would only incite gossiping and rumors from malevo-lent neighbors and fellow villagers, my hosts increasingly chose less exposed secondary streets and alleys to move with me through Angur Bagh, instead of the main bazaar road that we had used so many times. Over the years, between 2010 and 2013, I was introduced to fewer and fewer friends and relatives in the shops and gathering spots of the village. Associating with a stranger—worse, a non-Muslim foreigner—was considered by many with suspicion, and frowned upon, bringing possibly negative social consequences for the hosts. With the overall situation of the ongoing insurgency against the Afghan government and the foreign troops deteriorating, the possibility to participate more actively in the life of the household and the village that hosted me became more and more problematic. Over time I witnessed the growing uneasiness and even discomfort of my hosts for my being in the vil-lage. Nor did I fail to notice the increasing number of times when my request to join them in the village was refused with shaky and embarrassed excuses, that continued for several weeks. I could not visit the village any more after the killing of a US Army sergeant in Angur Bagh in late March 2013 at the

---

[1] In all fairness, I will say that the current limit in the display of melmastia, which I have personally experienced, was one of the aspects of social and cultural change that many among my informants regretted, as a byproduct (among many others, as we have so far seen) of the long decades of con-tinuous conflict. Many elders complained that the present deteriorating security situation, in con-junction with years of social chaos and impending danger, achieved the result of "pragmatizing" the choices that a family, or a lineage, would make, severely curtailing the performance of what had erst-while been a pivotal feature of Pashtun society and culture (hospitality).

hands of a local sixteen-year-old boy who stabbed the soldier in the neck, before disappearing on a motorcycle, which allegedly was waiting to whisk him away from the area. The incident had a strong emotional impact on the villagers and ended my visits to Sama Khel and Angur Bagh altogether. When I met friends from the village after the killing of the army sergeant, I had to do so in a concealed, almost secret fashion, away from the village, in a bazaar town along the Peshawar-Jalalabad road several kilometers removed from Sama Khel. There, very few people would be able to easily recognize my friends, but even so we chose an out-of-sight compound away from the gaze of passersby.

I went to Sama Khel for the first time in summer 2010 thanks to an improbable encounter with unexpected ramifications. An Afghan friend of mine in Jalalabad, a sharp young man who was studying English at Nangarhar University, put me in contact with an American man working for a foreign NGO in town. He was living in Jalalabad with his wife and his two toddlers, after having moved from Kandahar for security reasons. We met, together with my wife, in his NGO office, and then one more time at his house in Jalalabad. After listening to the description of our research, he said he knew somebody who might be interested in talking to us, and showing us his village. He made a quick phone call in Pashto to give the person our names and general information. He then returned to his conversation with us. I did not think much of the whole matter until two days later when someone knocked on our hotel door in Jalalabad at 7 a.m. Quickly putting on some clothes, I opened the door and walked out, carefully obstructing the view onto the interior of the room where my wife was still half-asleep. Three men introduced themselves to me in Pashto, very warmly, as Hajji Zia, the man with whom our American NGO worker friend had talked on the phone about us, and two friends of his. Zia told me quite curtly that we should immediately follow him to his village, where we would be his guests for the day. In such an unexpected way began my long frequentation of the village of Sama Khel, which lasted until I left Afghanistan in June 2013.

Hajji Zia was a man of about thirty years of age in 2010 and the grandchild of a brother of the leader of the Waraki lineage, Hajji Wahidullah. After spending time in Hajji Zia's private hujra in Sama Khel, during my first visit to the village I was also brought to see Hajji Wahidullah, in the old village hujra. Alongside him I met for the first time many of the male members of the lineage, whom I was to meet often in the subsequent months and years. A man in his mid-seventies, Hajji Wahidullah was a soft-spoken, slender, and meditative personage, who measured his words yet often expressed curiosity about me and my personal story. He was obviously well aware of his social position, and wore it with gravitas. I was informed by a relative that there had always been two *gunduna* (two sides) among the Shinwaris in the area: the ones loyal to Hajji Wahidullah and his khel, the others loyal to another elder, from a different khel. At the moment, the other figure proved more powerful than Hajji Wahidullah and had been elected to the Afghan Wolesi Jirga (the lower house of the Afghan parliament) as the national representative for the Shinwari *qawm*. "Things change," this relative told me, "next time it will be Hajji Wahidullah's turn to go the Wolesi Jirga," ostensibly not much worried by the political victory scored by the opposite side.

The "perfect" Pashtun

One of the grandchildren of a brother of Hajji Wahidullah was a man called Rahmat, of about thirty-two years of age in 2013. He represents, as we will see, a radically different figure from the ones I have discussed so far. While Rohullah, Umar, and Baryalay all, in their own ways, protested and rejected at least some aspect of Pashtun idioms of masculinity, as well as ways of displaying their "Pashtun-ness," Rahmat presented himself as perfectly adjusted to his sociocultural milieu. This at least was the public image that he conveyed to others (including me), as well as to those close to him on a daily basis. Indeed Rahmat seemed to live a seamlessly well-adjusted life in his own paternal village, and to enjoy the customs and the routines of his village life. But Rahmat's life was in reality far from being so clear-cut and idyllic.

Rahmat was one of the people I met first in Sama Khel's main hujra. I used to see him there only late in the afternoons, because he worked for an Afghan company that subcontracted the supply of flour to the US Army's Forward Operating Base that lay near the village, and he had to work long hours. After the sun set and everybody had moved from the outside *dera* (a shaded area with traditional *kat* to sit and lie down on) to the indoor hujra proper, Rahmat would show up at the end of his shift at the army base. Sometimes he would have night shifts, and would not participate in the hujra evening meetings for two or three days, going back to sleep at his family's house (which I have never visited). Nevertheless, anytime he joined the others, he showed and maintained a very dignified and distinguished appearance. His *shalwar kamiz* (the traditional dress) was always spotlessly clean, his beard well-trimmed, and his *Charsaddi chaplay* (a type of elegant leather sandals from Peshawar) were always clean and free from dust. He often wore a cologne that spread through the room when he entered, and he checked the state of his beard and moustaches on a regular basis through the small mirror incorporated in the round metal container that held his *naswar* (the slightly intoxicating powder that most men put between their teeth and gums, as in the US is common to do with chewing tobacco).

He was also one of only two villagers able to speak English, which he used to communicate with me, the other one being one of his young cousins, a seventeen-year-old boy called Kamran, with whom I also developed a good relationship. The fact that Rahmat had a good command of the language facilitated his being hired at the army base, where he managed to befriend several US servicemen and officers. According to the various letters of recommendation that his US military employers and supervisors wrote for him, he was a hardworking, serious, reliable individual, who impressed his co-workers with his work ethic. Rahmat was understandably quite proud of his recommendation letters, which he showed and copied for me. He also declared the intention to use them to get some sort of authorization to move to the United States, or somewhere else, in a near future. His dream has yet to materialize.

Rahmat struck me quite soon as a mild-tempered person, very much aware of, and compliant with, the unwritten norms of etiquette that govern Pashtun male gatherings. Accordingly, he never showed himself too vocal, or overwhelming in his participation in communal conversations. He was, as opposed to many who regularly attended the evening meetings in the main hujra, still considered young, and as such he had to maintain a composure and respectfulness toward the older family members that prevented him from appearing too daring or vociferous. His words seemed to be always well-chosen and pondered, for which, it seemed to me, he received in return attention and consideration from his relatives. His younger cousins (among whom was Kamran) held him in great esteem, as an upstanding, honest, and hardworking Pashtun *nar* (a manly man). His proficiency in English made him spontaneously gravitate toward me more than others did in the beginning. His acquaintance with many other Americans, at the base where he worked, made me perhaps appear in his eyes less of an exotic and suspect individual, to be kept possibly at arm's length rather than approached and befriended. This latter attitude was shared by some among the people who frequented the hujra. However, because Hajji Wahidullah, the living leader of the whole lineage, had always been much more accepting and friendly, this was sufficient to override all other considerations and opinions. Still, Rahmat's approach to me in the presence of others remained circumspect and cautious. But he wanted to know about my life in the West, and was ready to answer questions about his own life in the village.

The first conversations I had with him were in summer 2010, during the evenings spent in the village hujra. These were casual, and part of public discussions. It was only during my last stint of fieldwork in Afghanistan, in 2012 and 2013, that I asked him if he would sit with me, in a private place, and talk about his own life, without external interference. He agreed, and we eventually managed four sessions alone together in the house I was renting in Jalalabad. I guess his inquisitive and curious mind saw in my proposal also a good opportunity to answer some of his questions about the Western world (of which I was obviously seen as a prototypical specimen, as often happened during my fieldwork, without much discrimination being made between Europe and the United States). I also thought I would represent one more possibility he would have in the future to hopefully receive some help in trying to leave Afghanistan—illegally, or on a special visa, for which purpose he would need a Western "sponsor," or at least a guarantor.

Rahmat's sharpness first struck me. Our first conversations in the family hujra revolved around issues that would emerge as a corollary to the more poignant discussions taking place around us. For example, a few days before one of my visits to Sama Khel in 2010, two shops that Rahmat's father owned in the Angur Bagh bazaar were burned down. Blame was assigned to the members of the so-called 24 families (see below), who hoped to take a valuable piece of land that belonged to Rahmat's father. As the eldest son of his father, Rahmat was responsible to deal with the potentially explosive situation. He explained to me what happened, by translating the discussion in the hujra (we spoke English, as at that time my Pashto was not good enough to understand the heated hujra conversations):

You see, we Waraki [the name of his paternal lineage] are Shinwaris, we have been here for a long time, we came from a district near the mountains. The village down the bazaar of Angur Bagh, which is called Shergar, is inhabited by those people that we call the "24 families." They came from outside the area some time ago, maybe two or three generations ago, at different times. They are not Shinwaris, some of them were not even Pashtuns, they were

Farsiwaan [those who speak Persian]. *Now they call themselves Shinwaris, and some people start to forget that they came from outside . . . but we don't forget . . . they were people who abandoned the place where they were born, and moved somewhere else . . . what kind of person abandons the place of their birth? Why would you ever abandon the place where you were born? You do so if you are not respectable, if you are greedy, if you want more than you will ever have in your home village . . . no one should trust somebody who abandons the place where they were born. And in fact, these 24 families, as we call these people, are not to be trusted. They are after land, they are after money, they make problems and cannot be trusted. . . .*

ANDREA—*They are all bad people?*

RAHMAT—*Well, no, I mean . . . obviously there are some good people among them, but in general you'd better not trust them . . . when you deal with them you have to remember they are from the 24 families. . . . For example, some among them decided they needed this plot of land in the bazaar that my father owns, even though we are not using the shops that are there at the moment. They want the land, so they started with burning down the shops. There was a confrontation in the bazaar the other day between some from my family and some from those people. Fortunately nothing happened.*

ANDREA—*So, now what is going to happen? Are you planning to take revenge against those who burned down the shops?*

RAHMAT—*No, I don't want revenge now. It's not wise to start any big problem for us now. We have the right documentation to show to the authorities, and with that we went to the district governor. We showed him that the piece of land belongs rightfully to my family. Now we will see what happens. Hopefully the police will help us, so that this thing does not happen again.*

ANDREA—*Nobody in your family is telling you that you should take revenge? Nobody is giving you* peghor *about it?* [peghor is the moral condemnation one receives for failing to defend with ghairat one's compromised honor].

RAHMAT—*Yes, some have come to me in private and told me that the honor of our family has been insulted by this attack, and that we should respond in the same way, to demonstrate that we can defend our rights, and what is ours. But I believe that it would not be smart at this time to do so, and that we can start with other ways. There is always time for revenge. The elders of the family have agreed with me, and we are going by steps. First we have gone to the district governor . . . we'll see what happens.*

During this conversation, Rahmat appeared extremely sure of himself, calm and convinced of the rightfulness of his choice of action. While showing himself as a person deeply immersed in, and cognizant of the way in which certain things are dealt with traditionally, he equally proved to be pragmatic and level-headed enough to avoid getting sucked in by the most visceral and sanguinary aspect of pukhto (the Pashtun customs): revenge. Rash reactions, although in theory in line with a strict understanding of pukhto, have the power of starting vicious circles, lasting for generations of bloody retaliations. We have seen in my previous informants' life histories how easily incidents like these can get out of hand and snowball uncontrollably. Rahmat gave me, and presumably his older family members, the impression of being a wise and mature person in control of a delicate situation, with a reasonable plan of action for the future. And he also presented an image of himself as a man firmly rooted in his traditional social environment and cultural background.

Nonetheless, his tirade against the members of what he calls the "24 families" was unexpected. At least to me, it sounded extremely passionate in its moral reproach. The ties to one's own family and place of birth and belonging have always been known to be a crucial aspect of Pashtuns' Weltanschauung. The (somewhat inherently nostalgic) public narrative of the importance and emotional power of one's own Heimat (the motherland) had always seemed to me to be the ideal, utopian scenario against which reality would necessarily clash, and force people to (supposedly unwittingly) adjust and make virtue out of necessity. Indeed, in spite of the stereotypical image of living in a rural and static world, most Pashtun families historically produced at least one migrant laborer, who stayed abroad for many years, and often came back a changed, although more affluent man. Historian Robert Nichols meticulously documented 230 years of Pashtun migration routes and their vicissitudes, from what today is Afghanistan and Pakistan to India, the Persian Gulf states, East Africa, and even beyond. In some cases, entire villages relocated to India, and their inhabitants became an accepted element of the Indian social landscape (Nichols 2008). Likewise, during my own fieldwork research, I rarely encountered individuals who had not experienced internal or external displacement, labor migration, or some other sort of disruptive event that had forced them to leave their home village or town, at least temporarily. So I had never encountered before anyone among my informants who expressed so indignantly their moral reproach at those who had to leave, or relocate themselves. Thus, Rahmat's highly emotional words surprised me, although he himself later proved to be far from applying such standards to himself.

An additional element of Rahmat's "traditional" thinking was brought home to me during a different visit to his village in 2011, when he disparaged the status of one of his acquired relatives, Imran Pacha. Imran was a member of a pacha family, living in a village close to Sama Khel, and his sister had recently married one of Rahmat's cousins. Although Rahmat praised Imran for being an honest and respectable man, he harbored a barely disguised resentment toward Imran's family and its social position as pacha. Discussing the recent marriage of his cousin with Imran's sister, he said:

They used to be very well respected. His father and grandfather used to make amulets for the people of the area, and help people with solving disputes that would arise among them. Imran does not make amulets anymore, he has become more a political figure here. He still gives opinions and works as a jirgamaar when asked to [the jirgamaar is one of the members of the Jirga, the council of elders appointed by two parties in order to solve their ongoing conflict]. *But his brothers have started to take advantage of the wealth that their father and grandfather acquired. You see, people would reward their amulets, counseling, and opinions with gifts, sometimes even land. They have accumulated a good deal of property, and now Imran's brothers have become politically more influential.*

ANDREA—*But people consider them Pashtuns, like any other, right?*

RAHMAT—*Yes, they generally do, but the truth is that they came from Iraq, at the time of Abdur Rahman* [end of the nineteenth century]. *They are Arabs, really, I consider them Arabs. They are one of the 24 families who came from outside. They became powerful and respected because they were pacha, but a lot of what they have now comes from gifts of the Pashtuns who live here, they did not have to work and sweat for what they have now. One of Imran's brothers is among those who burned down the shops we had in the bazaar . . . he is with them.*

ANDREA—*Is there any difference between the lifestyle they conduct, and yours* [as Pashtuns]?

RAHMAT—*No, in reality you cannot tell the difference now between one of us and one of them. The only difference maybe is that, when you have to start getting information about the possibility of a marriage, they begin by sending out the women, while we begin by sending out the men. Our women come later in giving their opinion. . . .* [long pause] *What about you, in*

*the West? I know that women do a lot of things in the West, they get polit-
ical positions, they have a lot of freedom in daily life . . . what do you think
about it?*

At this point of the conversation, it seemed to me appropriate to embark
in a lengthy and tactful monologue about the reasons why in "the West" we
had (in theory, that is) parity and equality of treatment and opportunities between men and women, trying to explain that, while I did consider that state
of facts right for our cultural environment, I also did not consider Pashtuns'
social arrangements necessarily "wrong" *tout court*, but rather as the inevitable outcome of different historical and cultural developments. My sermon,
however, did not seem to convince Rahmat very much. He interrupted me at
a certain point, and declared:

*I don't know . . . to me women are just weak, both physically and men-
tally. . . . I mean, it's obvious, you can see it every day, they cannot perform all
the duties and cannot sustain all the stresses that men have to bear . . . that's
why they have to stay in the house . . . what would they do without men, outside? They would be unable to function without men. . . .*
ANDREA—*But what about all those women who do men's jobs in the West? If
women were really so much weaker, how could they do all those things they
do in the West, and be successful? There are women in the government, in
the schools, in the military, in private companies. . . . And what about giving
birth? Don't you think that giving birth is a very stressful and difficult thing?*
RAHMAT—*I think in the West there is too much freedom . . . everybody can do
whatever they want, there are no rules, no order . . . look what happened
with AIDS: everybody is free to have sex with anybody else, there are no
moral rules, and big problems like AIDS, divorce, and other things become
huge problems. . . . I think it's a wrong way to live. . . .*

The narrative about group purity (metonymically embodied by the concept of Heimat), and the supposedly unique virtues of the in-group (in this
case, shrinking considerably to the closer boundaries of the sub-group
Shinwari), are idioms that we will encounter again, and have nothing particularly notable in this context. Rahmat, however, expressed these concerns
and beliefs in a very matter-of-fact way, without inducing me to think that
he might have been staging a performance for the sake of capturing the interest of the nosy, and naïve, foreign anthropologist. Rahmat did not attempt

to present a sanitized view of Pashtun reality for "Western" consumption. The sense of "embarrassment" and uneasiness that I perceived in some other informants about what they thought were the "bad things" about their society, to be kept away from the understanding of a Western observer, was apparently absent in Rahmat's self-presentation to me. Speaking to me in the hujra in English, he was not subject to the others' criticism (being the only one who could speak and understand English). That rendered our conversations de facto private exchanges, presumably reflective of his "authentic" dispositions and his own feelings. He was at the same time curious about my world (which in his mind was "the West"), and did not seem to share the belief of other informants that "the West" was in some way socially and culturally superior.

This impression on my part I believe was corroborated by the uncompromising stance that he expressed about women. He knew that his views on women would most likely not gain much sympathy from a Western interlocutor. He had had, as we will see, enough experience with foreigners and outsiders to realize that. Yet he did not want to impress me, or to sound more "correct" to my ears, and simply stated and defended what to him seemed not only obvious, but even undeniable facts. His take on women is certainly not isolated among Pashtuns in Afghanistan, and as such does not represent either a surprise, or great news per se. It was the firmness and openness with which he asserted his convictions that left me somewhat surprised. I was used to more "diplomacy," so to speak, from many of my informants, stemming from the underlying concern that they would in the end give a bad "image" of themselves, and of Pashtuns in general, to a Western interlocutor. Not that their deep beliefs and convictions would not emerge, sooner or later, in other, indirect ways—through open behavior, overheard comments in Pashto, seemingly innocent jokes, and moral judgments on other people's vicissitudes. Yet I always took into consideration this degree of formal "diplomatic talk" when approaching someone I did not know well enough, and corrected for it accordingly. With Rahmat it seemed that I could skip the diplomatic talk stage, and take what he reported to me about himself as a more direct reflection of his thoughts and feelings. We will see that this peculiarity in him, and in our relationship, became permeated by interesting meanings, in relation to the kind of person he seemed to be—or at least, that he presented to me. The "traditional" shape that his character started to take in my eyes, and the fact that he chose to remain tied to a village life and environment, proved to be aspects that strictly related to each other.

These stimulating occasional contacts with Rahmat, which I had several times in 2010 and 2011, brought me to the decision of asking him, in 2012, whether he would sit with me and talk more in depth about himself, his life's trajectory, and future plans. He accepted, and we managed to arrange the first session (we had four in total, of about two hours each) during the last section of my fieldwork, in 2013. By then, the situation in the district where Sama Khel was located had already deteriorated so much that he did not want me to go to his village, and offered to come to my house in Jalalabad. Though I had hoped to talk to him in his own natural environment, still meeting in Jalalabad had its advantages, ensuring that there would not be any external interference, or pressure, during our conversations.

Complicating the portrait

Rahmat Khan Waraki was born in Sama Khel around 1980 and moved to Pakistan with all his family when he was about four years old.

We all moved to Peshawar a few years after the war against the Russians started. We did not go to a camp though [a refugee camp]. *We went to live with a relative of ours who lived outside Peshawar, in Jamrud. Back in the time of Abdur Rahman* [1880–1902], *one man from the Waraki family married into an Afridi family on the other side of the border* [Afridi is a different sub-group of the Pashtun people, mostly based in Pakistan], *and they started a family there. They still live in the same place, and we have always had friendly relationships with these people over the years. They agreed to give us a house in which to live in the area where they too were living. I was put in a school for refugees, and studied there until third grade. Then we moved to Khyber Agency* [one of the seven Tribal Agencies of Pakistan], *where we bought land and built a house. I went to school there until eighth grade. I was playing cricket at that time, and I was really good at it. . . . I was a promising player. . . . I stopped going to school, and started working in a factory to pay for cricket expenses. I did that for about three years. The coach of my team told me that I should try for the junior Pakistani national team, but I refused, because I would have to get Pakistani citizen-ship, and I did not want to do that. That's where I learned English . . . you know, we were going around other provinces to play, and everybody would speak English. . . . I learned good English because I thought that I could do*

something with cricket, and English was necessary . . . it did not go through,
maybe it was my fault. . . . I was a good player though. . . .

ANDREA—*Why did you not continue with cricket?*

RAHMAT—*Well, there was this cousin of mine living with us in Khyber Agency,*
we were thinking of how to make some money. . . . I was not happy there,
I wanted to change and get something for myself. . . . [long pause] Also, I was
the eldest son on my family, I had two younger brothers, five and eight years
younger than I was, I felt that I had responsibilities toward them. . . . Me
and this cousin of mine, we saw that some easy money could be done with
opium. We decided that we would go back to Sama Khel and start culti-
vating opium poppy. We did not need much, we had only two jiribs [slightly
less than one acre] available for us, and we used it all for poppy. I sent the
money back to Pakistan to my family.

ANDREA—*Did you go with anybody else back to Sama Khel?*

RAHMAT—*No, I was the only one from my close family, I was the only one*
who went back, with this far cousin of mine. We did that for several years,
until maybe the third year of Karzai's government [2004]. There was still
somebody from my extended family in Sama Khel, so I lived with them. We
were making good money with the opium poppies. We were able to develop
from there, and started collaborating with people that managed a labora-
tory that produced heroin from opium. We would work inside the labora-
tory making that stuff. . . . [long pause] It was not a good period . . . a lot of
people coming back from Pakistan, they brought back here bad habits, bad
behaviors . . . bad things started happening, violent stuff, the situation got
worse . . . money changes a lot of things, you know, and also people changed
how they behaved . . . there was the war, you know, people were affected
by the war, even though they did not fight personally. Some did fight, some
did not, but the environment had a lot of violence in it . . . and then there
were the people who came back from Pakistan, they brought back different
values, different ideas. . . .

ANDREA—*You as well were among those who came back from Pakistan. . . .*

RAHMAT—*Yes, I know, but for me it was a little different. . . . I had lived in a*
house with my family, away from the city [Peshawar], and away from the
camps, for me it was like living in Afghanistan somehow, so I did not get
influenced very much by certain things. . . .

ANDREA—*Certain things like what?*

RAHMAT—*Well, you know, the life of the city, buying things, wanting always*
more things, more money, never being happy with what you have in your

home place . . . those people were not happy anymore with the qala [the mud-brick house], *they wanted a bigger house, made of bricks and cement, a nice car . . . and also, I could see the people I worked with in the laboratory* [for heroin], *they would behave badly, there was a lot of violence involved with the job. . . .* [long pause] *you know, when someone starts doing something wrong, and then others follow them, in the beginning people notice, and say something, but if the behavior continues, and more people behave like that, well, then in the end it becomes normal, you know, people do not notice anymore, they think it is normal to do that . . . that's how things get bad.*

ANDREA—*Well, there are a lot of people who came back from Pakistan . . . are they all so bad?*

RAHMAT—*No, of course not, there were also good ideas that they brought back, like education . . . in Pakistan it is very important, you know, people know that it is important to send your children to school . . . here in Afghanistan it was not like this, but now, with all the Afghans that were born in Pakistan and went to school there, now also in Afghanistan we understand that it is important to send children to school . . . also girls . . . maybe until fifth or sixth grade. . . .*

ANDREA—*What about the heroin laboratory?*

RAHMAT—*Well, after my parents and close family came back from Pakistan I stopped working there. I found some jobs in Jalalabad and Kabul . . . it did not last long, I did not like the city, I came back here to Sama Khel.*

ANDREA—*Why didn't you like the city?*

RAHMAT—*I was alone, I had strange jobs, only for short periods of time, like construction. . . . I had nobody in the city, I had to share a room with other people I did not know, they were strangers, I did not know where they were coming from* [meaning that he did not know their background and family history]. . . . [long pause] *Here it's different, you know . . . when I worked for the flour company at the Army base* [he had lost his job a few months earlier, in late 2012], *it was hard, we worked long hours, seven days a week, sometimes in the evening or at night, but it was ok, I like to work . . . after work I would come back home, there was the hujra, there was Hajji Wahidullah* [the current leader of the lineage, brother of his grandfather], *there were all my relatives here, we would spend the evening together, talking, drinking tea, listening to the news on the radio* [from Voice of America Pashto]. . . . *I liked it, it was where I wanted to be. And even now that I don't have a job, I can go down the bazaar where we have a few shops,*

I stay with my friends . . . we go to Marko bazaar [a few kilometers away], *we meet other friends. . . . I wish I had a job, but now at least I am with people I know, that I am happy with. . . .*

Rahmat had a comparably quiet childhood and early adolescence. While he had to leave his home village because of the war, he was presumably spared the worst of it and the fate of many other Afghans, such as life growing up in a refugee camp. By his own admission, he managed to recreate the familial village environment in Pakistan, or at least he felt the sojourn in Pakistan was not overly dissimilar from what he had previously experienced in Afghanistan. Yet something was not completely right. He did not finish school in Pakistan, in spite of the fact that by then his family had acquired their own land and built a legitimate house. An interesting aspect of his narrative is the fact that he speaks of the impact of the war and the atmosphere of increased violence it created, but as affecting others, and without putting himself in the same category. As previous psychological research has indicated (Charles Lindholm, unpublished data), it is often easier for respondents to attribute certain feelings, actions, or pragmatic choices, culturally considered culturally negative or reproachable, to others within one's own community, rather than to oneself. I had the impression that Rahmat was doing this. He did not give me any specific reason for his choice of leaving the Pakistani home to go cultivate opium poppy in Afghanistan with his cousin, while he underscored those "bad behaviors" that he perceived *other* Afghan refugees returning from Pakistan were bringing back with them as "social remittances" (Levitt 1998, Levitt and Lamba-Nieves 2011). When I pointed out that he was also a returnee, he dismissed the comparison by saying that his living environment in Pakistan had not produced such detrimental social consequences. However, he was clearly not particularly proud of his involvement in activities that he knew were both illegal, and condemned as immoral by Islam.

The trope of having lived in a familial, friendlier environment during his childhood in Pakistan, however, is used by him to corroborate a different interpretation of the situation. He was the eldest son, and like any other Pashtun young man in his position, he was the one expected to sacrifice his private strivings and ambitions (like playing cricket professionally) for the sake of his younger brothers' well-being. In taking up the role of the responsible provider, he was fulfilling a cultural norm, but in a way that violated more fundamental principles, causing a moral conflict. The opportunity to engage in drug production and trade was provided by the social and economic disruption brought about by the war in the first place. Rahmat never wanted to delve into the details of the period wherein he worked for the laboratory that processed heroin from opium. He mentioned the "bad behaviors" and the pervasive violence that were associated with such occupation, leaving to the imagination what these were. Again, he speaks in the third person (plural) when referring to the criminal activities that the laboratory and its staff were engaged in. He distances himself from what went on in and around the laboratory, attributing whatever responsibility to the "others" who worked with him.

However, his detailed and insightful description of how a deviant behavior becomes slowly routinized and institutionalized within a small community is remarkable. It felt to me that he could have acquired that intimate understanding of such a crucial social phenomenon only by active participation in it. I believe Rahmat did not want me to associate him with practices that he knew I would have probably not considered favorably, and which, more important, he himself judged "wrong" at the moment of our conversation. And in fact, his past spent working with opium and heroin appeared alien to the man I knew since 2010, who was a family member respected for integrity, piety, and rectitude. His virtues of tolerance alongside firmness and resoluteness were praised to me by his younger nephews. The person that I saw interacting in Sama Khel with family members and friends did certainly adhere to this public image. Yet the events that took place during the years in Sama Khel, spent between opium and heroin, did not. They did not fall into place with the self-presentation he was displaying publicly, and, above all, with the self-image that he seemed to maintain of himself while talking to me.

Referring again to the model suggested by Sandler and Rosenblatt (1962), we might say that his overarching self-representation, seemingly coherent and not fragmented during the time I knew him, managed to accommodate differing self-images (or states of subjectivity, selves) that had to serve different psychological purposes, and provide a psychically functional adjustment to life contingencies that were certainly not ideal. The moral standards that he verbally expressed and behaviorally displayed with me would not condone his past achievements as a heroin processer. The self-image and subjective states that allowed him to engage in such activities must have been disavowed over time (at least partially), and replaced with different ones, more congruous with a changed environment, as well as shifted private wishes, hopes, and social opportunities. On the one hand, contradictions between the two self-images and subjectivities (one past, the other present) did indeed emerge during his recounting of the events, and in the choice of not fully disclosing his real degree of participation in those activities. On the other hand, the overarching coherence in his broader self-representation was, I believe, indicated by the very fact that he was able to articulate verbally, and hence consciously acknowledge, the existence of these same contradictory self-images (and in so doing, bringing to awareness the conflict between existing self-images, or states of subjectivity, which turns the dissociation between them into the ground for psychological functionality and growth. See Bromberg 1996).

That, at the time when our conversations took place, Rahmat inhabited a different self-image than the one he had during his stint in the heroin laboratory (that, in other words, a different state of subjectivity was predominant in him), more adherent to a culturally normative set of moral standards proper to a respectable and honorable Pashtun man, illustrated his specific discourse about ghairat and honor during one of our private sessions:

Ghairat is based on honesty [ghairat de diyanat pure aralari]. Ghairat is that thing that makes you take action when you see injustice [zulm, also cruelty, abuse], when you see someone doing something that is not in accordance with religion, customs, and the people's values [arzakhtuna]. A ghairatman [a person with ghairat] will never ask for money to do what he thinks is best to protect these values. He will never complain about the hardships that he has to suffer in order to do something honorable. He will not go behind the back of somebody to talk badly of him, he will tell him to his face what he thinks [peghor].

ANDREA—Is this how you behave with other people?

RAHMAT—Well, you know me, you saw me in the village, don't I do this? You saw that this is what I do in my life. . . . This is what I believe right to do. . . . There are fewer and fewer people like this today, as opposed to the past. . . . I told you, I think this has to do with the war, and the people who came back from Pakistan. I think the best thing for a man is when he comes back from work at night and he can sit in his hujra with his family members. I was happy when I did this in Sama Khel after work.

ANDREA—There are many people who do not like the life in the village, this life you are talking about . . . they say there are too many problems in the village, too much fighting, too much conflict. . . .

RAHMAT—*Yes, I know, this is true, there are many fights and conflicts* [lanjay] *in the village . . . but you have to adjust, you have to take what comes, and not complain. . . . You have to be honest . . . if you are honest* [spin, literally "white," a white, pure heart, in this case] *you can survive anything. I know there are problems in the village, but it is still better than life elsewhere, like in the city.*

While this narrative might seem too ideally coterminous with the culturally shared and accepted prototypical character of an honorable Pashtun man, still Rahmat's family and friends regularly used to describe him in this way. Furthermore, my personal experience with him was consistent with it. It was not uncommon, during my fieldwork, to hear an informant praise and support the culturally normative version of how a man should act, but not see him live up to it. By contrast, Rahmat was the only person that I knew well who never asked me for any "gift" or favor, who never expressed or complained about his state of need with me, and never raised suspicions of being motivated by some underlying ulterior motive. His "heart," as far as my experience went, looked indeed "white." At the time of our conversations, he had apparently chosen to identify positively with the ideal prototype of an honorable Pashtun man, the embodiment of a Weberian ideal type of Pashtun "hegemonic masculinity" (Connell 1987, 1995; Demetriou, 2001, Inhorn, 2012). In this sense, Rahmat seems to present a case where the cultural representational world and personal self-image are strongly congruent.

The representational world produced by a cultural milieu interplays and may profoundly inform the modalities of functioning and the symbolic contents of the private representational world that individuals maintain for themselves. Rahmat seems to have significantly strayed away from a behavioral conformity with the culturally accepted moral standards for a respectable and pious Pashtun man when he was involved in criminal activities, but later returned to a model example of a *nar* (a self-reassuring, and possibly socially legitimizing position). Rahmat embraced this culturally hegemonic representation of masculinity almost with enthusiasm. Such overlapping between the cultural representational world, and the private representational world in Rahmat, must be considered, I believe, functional to the maintenance of his psychic equilibrium in this life conjuncture (an example, I would argue, of how pragmatically the individual is able to tap from unconscious emotional processed material, and render it meaningful in a given life conjuncture—what Wilfred Bion calls "learning from experience." Bion 1962).

It must be added here that, as it had happened to Rohullah in his own Paktia village, R. W. Connell's paradigm for "hegemonic masculinity" adapts well to the sociocultural environment of Rahmat's village. In a very traditional and "conservative" area of Nangarhar province, such as Angur Bagh district, there can certainly be detected, by and large, a univocal model for a dominant (and appropriate) performance of masculine behavior that rests squarely on the attribute of ghairat, and on values such as izzat, sharm, and parda (the segregation of the sexes). A set of diverse, and equally hegemonic *masculinities* (see Demetriou 2001)—the "luxury" that Rohullah found in Peshawar during his last two years of high school, and that Baryalay was able to take advantage of in his village, given his "saintly" genealogical background—was for Rahmat unavailable in his home village Sama Khel. Rahmat would not settle for a subordinate, painful type of masculinity, of the kind that the young Rohullah had to embody in the first years of his life in the Paktia province village.

The strength of interconnectivity

The overlap between the two representational worlds, between these two aspects of his subjectivity, certainly did not take place without difficulties.

A short time before going back to Afghanistan with the rest of the family, my father told me that I had to get married. I think he was getting worried for the things that I was doing in Afghanistan. I was still in Sama Khel at that time, working on opium with my cousin. I did not want to get married, I was not thinking about it. My father and my uncle told me that I should marry this cousin of mine, that they had already decided. I did not know this girl. She was living in Pakistan, close to my family, but I had never seen her, I did not know her. I told them that I did not want to do it. My father insisted that he had already promised my uncle that I would marry his daughter, and that it would be a great dishonor and shame if I did not do so. He said that he needed to have a good relation with this uncle of mine . . . he was not his own brother, he was an uncle of his really. . . .

ANDREA—*So, you accepted what your father told you. . . .*

RAHMAT—*Well, actually, no, I don't know if I would have done it if my mother had not been so sad. . . .*

ANDREA—*What do you mean?*

RAHMAT—*Yes, it was not my father that convinced me, it was my mother . . . it was my mother's words that convinced me that I should do as my father was telling me to do. My mother's words were stronger than my father's authority [waak]. She was extremely sad, and she was always crying and lamenting with me that I would destroy the reputation of the family by not marrying the girl. . . . She was very sad. . . . I could not see my mother like that, so I said to my father that I would do it. Before I said yes, he also tried to convince me by saying that after marrying this girl, he would give me permission to sell a piece of our land to pay for a second marriage with a girl I like . . . he said that's how it went for him, and he did not regret it. . . .*

ANDREA—*Is that what you are going to do?*

RAHMAT—*No, no second wife, it's too many problems. . . . I don't care, I am fine with my life like it is now. My wife and I have a good relationship. You know, that does not happen very often. . . . I mean, it is common for a husband and a wife to have a bad relationship when the marriage is arranged. . . . They will have hard feelings against each other, they will try to get revenge on the other. . . .*

ANDREA—*Wait, why do you say revenge? [badal]*

RAHMAT—*Yes, revenge, because both are angry, both are upset because they have to stay with a person they don't like, and cannot have their own life, they could not choose a person that they liked . . . sometimes, you know, people have somebody they would like to marry, but they can't . . . so husband and wife fight all the time, they are always angry, and the sweetest time of their lives is wasted this way. Then you see that they calm down when they get old, and they stop fighting because they are too tired to keep fighting . . . but at that point it's too late, their young years are gone, and they become old people. . . .*

ANDREA—*This is not happening with your marriage?*

RAHMAT—*No, I told you, my wife and I have a good relationship. I don't want any revenge on my wife. I don't feel that way. I try to be kind and respectful with people that do not deserve to be treated with violence. . . . My wife's feelings are important to me, I do not want to hurt them. . . . I mean, how could you hurt someone who loves your children more than you do?*

ANDREA—*Well, someone does, you said that it often happens here. . . .*

RAHMAT—*Yes, I know, I know, I see them. . . . I don't know where this aggressiveness* [ta'aruz] *comes from. . . . I guess from bad experiences, or from natural dispositions, and then I told you, arranged marriages bring bad blood between spouses. . . .* [long pause] *I don't want to hurt a weak person, I am afraid of that, it would not be honorable . . . but when some rights are in danger and must be defended, like your family's rights, or your family's honor, I am not scared even of the most dangerous actions. . . . When I see violence done to the powerful and the rich, I have no problem with that. . . .*

ANDREA—*Wait, why is that so?*

RAHMAT—*People who are very powerful and very rich around here, they did not make what they have with honesty and honor, they stole and committed* zulm *against those who were less powerful. . . . I don't care if someone does violence to a* kumandan [a militia leader], *many others have suffered because of him. . . . For example, let me tell you what happened about six months ago. The grandson of Hajji Wahidullah was kidnapped by a group of powerful people that had a problem with us about some land . . . they wanted to convince us into some deal. . . . They kept the boy for three days, then they released him. A few weeks later, me and a friend of mine, we were in his shop in Marko bazaar, and we saw two of the group who kidnapped the boy. We dropped everything, and we ran after them, we caught them, and we beat them right there, in the middle of the street. Some people stopped us after a while, and then the police arrived, but they did not do anything. They knew me, and they knew what had happened to the boy. We all went home. That's how it should work, you know, I did not do that for showing off, I did it because it was the right thing to do. . . .*

The fact that Rahmat was forced by his close family members to accept an arranged marriage with a girl he had never met is of course nothing exceptional in Afghanistan. We have seen already that Umar was able to describe with remarkable introspectiveness the feelings and emotions he went through while consenting to his family's wishes and adjusting his personal life in order to accept the inevitable. Rahmat was subject to the same process and likewise adjusted to the role he was assigned by his family. One interesting commonality between the two cases, all the more so due to the marked difference in social and educational backgrounds between the two informants' families, as well as between Rahmat's and Umar's personal profiles, was the prominent role of the mothers. Umar's mother was a strong and willful woman, who imposed herself on her son in order (according to

Umar's perceptions) to be able to exert a strong authority over her future daughter-in-law. Rahmat's mother induced his cooperation by playing the "compassion card" to elicit feelings of guilt into his son Rahmat, and push him into accepting the family's decision.

Psychologist Marwan Dwairy and his colleagues (Dwairy et al. 2013) detected a cross-cultural pattern of behavior in a recent study of several Muslim environments in the Middle East. Here, too, family "discipline" was enforced through a strategy they termed "compassion evoking" in a sample of Arab Israeli, Sunni Muslim Lebanese, and Arab Algerian families. Some of Dwairy's informants reported of their mothers that "[she] often tells me how much she and the family are suffering because of me." Whereas fathers, as Rahmat's father did, take often an authoritarian stance in the matter by demanding that their sons do something, mothers play on their son's feelings and sense of familial obligation. Whether this is done consciously and surreptitiously, or unconsciously, we cannot tell either from Dwairy's study, or from Umar's and Rahmat's accounts, and it surely varies in relation to the personal idiosyncrasies of the protagonists—although in Umar's case, his own impression of his mother's behavior clearly hinted at a consciously orchestrated plan on her part, in order to receive as a daughter-in-law an easily controllable and malleable girl.

Rahmat's mother's behavior, and Rahmat's acquiescent response to it, seem to replicate a cultural pattern seemingly found across the large geographical belt that stretches from North Africa and the Mediterranean, to the Middle East, up to South Asia (not only among Muslim social milieus). Now, how do we interpret, at a psychodynamic level, Rahmat's acquiescence and subsequent adjustment? It certainly represents a clue to his profound interrelatedness, or "connectivity" (Joseph 1999: 1–17) with his socially and emotionally "significant others" (his extended family), and to his interpersonal dependency on his social milieu. Such a condition, I believe, is not necessarily a sign of what Katherine Ewing calls "intrapsychic enmeshment," hinting at a negative and pernicious blurring of ego boundaries (Ewing 1990). I think that both Joseph and Ewing are correct in considering the culturally shaped interpersonal relatedness, and the profound consequences that it has on the individual's psychological dynamics, as a feature of certain cultural milieus (such as the Pashtun one) that should not be seen as either pathogenic per se, or indicating a lack of sense of individuation, personal autonomy, and agentic power (as is argued by many psychologists and some anthropologists; see, for instance, Markus and Kitayama 1991, Sharabi 1988, Minces 1982).

Ewing also maintains that "family members [who] fail to perceive each other in terms of their own unique attributes and needs, but instead fuse their perceptions with their own intrapsychic representation of conflictive internalized objects (projective transference distortion)" display a dangerous lack of intrapsychic differentiation between self and other (Ewing 1990: 138). This viewpoint will account for part of the spectrum of interpersonal relationships, but I believe in cases like Rahmat's it may not accord the due weight to deep-reaching cultural idioms and patterns for emotional engagement between members of an extended family (or even friends). One must also take into consideration the conscious and unconscious calculations toward the protection and enhancement of self-interest and personal gains, which are reached through conforming to those culturally enjoined patterns of behavior and feeling that are a significant part of any interaction between individuals. As the cross-cultural psychoanalyst Alan Roland suggested, cultural models and constraints affect the deepest layers of the human psyche (A. Roland, personal communication, March 2011). In cultural environments wherein interpersonal relatedness is at the center of the moral and emotional landscape of each individual, cases of supposed "intrapsychic enmeshment" might be otherwise explained, without recourse to pathologizing analytical categories and processes.

Rahmat, for example, I believe never failed to sense where his personal priorities were, even when he chose to accommodate his parents' wishes. He was clear in his mind where his personal choice, as if abstracted from a sociocultural context, would rest (i.e., in not getting married then, and in waiting to find another suitable girl). Yet no personal choice can really be fully abstracted from the surrounding context, and Rahmat seems to have perceived that as well. Rahmat somehow realized that the happiness and well-being of his mother (and of his family at large) was more important for him than the fulfillment of his own personal fantasy of marrying a more suitable girl, a fantasy that could not truly materialize in a real world made of social connections and cultural models (his contingent "social figuration," in the words of Norbert Elias. Elias 1994 [1968]: 211–215). The happiness of his mother meant for Rahmat *his own* happiness as well. In the context of his Pashtun cultural environment, Rahmat's *own* self-interest lies together with his family's interest. Emotions, "authentic" feelings, and hence the sense of autonomy and individuality, follow accordingly. There is no "psychic enmeshment" at the root of Rahmat's decision, I argue. Personal, private wishes get adjusted to, and become openly shaped by the cultural and social schemata of one's own environment. The new shape these wishes adopt becomes felt as one's own, and "authentically" so, just as much as one's disconnected-from-reality fantasies had been prior to the necessary reality-check. These two apparently conflicting registers of subjectivity, I would argue, belong to two different self-images, which the individual is able to construct and manage coherently, when in possession of a healthy psychic equilibrium. It appears clear in this sense how the notion of a bounded ego (or self, see Geertz 1984: 126), isolated and insulated from social and "cultural figurations" (paraphrasing Elias), should be revised.

And in fact, while talking to Rahmat, as it happened while talking with Umar about the same issues, I felt that they had never abdicated to their own "individuality," never "compromised" their ego boundaries by fully identifying with their mother's wishes. Heinz Kohut would treat this as an unhealthy dependence on the *self-object* that the mother's wishes represented for the individual, due to a poorly grounded self (Kohut 1972). In my view, it was not with his mother's wishes that Rahmat identified, renouncing to his own ones, but rather with the sociocultural idioms of his own environment. These idioms are not only those that pertain to the idea that the children of a family should follow the directions of their parents in their marriage plans. If it was so, Rahmat could have just as well obeyed his father in the first place. What I am rather referring here to is the specific role that mother has in the "family play" altogether. As we have seen with Dwairy's and his team's findings, there seems to be a neat "division of labor" within the family about who does what and how, in the relationship with one's own children. Adhering to a certain cultural template, whether consciously or unconsciously, means for Rahmat, as it did for Umar, to concede that particular role to mother and act accordingly. In so doing, Rahmat's own self-interest became coterminous with complying with those injunctions that stem from such particular mother's role. I believe that in this case the perceptive insights by relational psychoanalysts Malcolm Slavin and Daniel Kriegman may come to our help to understand Rahmat's inner dynamics. They write:

> In relation to the evolved need to maintain a sense of cohesion or integration in the experience of the self, states of relative fragmentation may be designed to occur when the maintenance of cohesiveness becomes secondary to the need to tap into the store of alternate identity elements because the current fit with the environment is poor. Although subjectively painful because they are experienced as "deficiencies" or "defects" in the structures that create or maintain a

sense of wholeness, the inner loss of a sense of wholeness signals the operation
of an evolved strategic process in which multiple self/other schemas are revived
and reanimated in an attempt to revise self structure or induce changes in the
environment. (Slavin and Kriegman 1992: 206)

There is no doubt that it was painful for Rahmat to abide by his mother's
and father's injunctions, and marry the girl they wanted him to marry, just as
it is very likely that he experienced a sense of fragmentation of his perceived
wholeness. Yet his self-image of the "good Pashtun man" was here conjured
by Rahmat to mend a "deficiency" he encountered in the external environment, and prioritize that against his self-image representing his more private
wishes, sometimes at odds with external reality. Both self-images, or subjective states, are Rahmat's own, so to speak—they are both "authentically"
embodying a piece of Rahmat's wholeness. What Rahmat is doing here (and
to this day, after years of marriage), is successfully "standing in the spaces," in
the words of Philip Bromberg:

Health is the ability to stand in the spaces between realities without losing
any of them—the capacity to feel like one self while being many. "Standing
in the spaces" is a shorthand way of describing a person's relative capacity to
make room at any given moment for the subjective reality that is not readily
containable by the self he experiences as "me" at that moment. (Bromberg
1996: 274)

In more general terms, such apparently contradictory conflation between
sociocultural expectations and self-interest, or perception of wholeness,
was resolved successfully by Rahmat through developing a set of emotional
inclinations and attachment priorities that he perceived positively as his own,
"authentic" to his own deep needs. In other words, Rahmat felt somehow that
he "authentically" cared more for his mother's well-being and peace of mind,
than for his own fantasies and wishes, which he deemed unrealistic within a
real-life social context. This I take to be a successful adaptation of one's emotional creative production, rather than the clue to any pathogenic instance of
"intrapsychic enmeshment." I believe that this process is a testimony to the
powerful role that sociocultural patterns and norms acquire in shaping the
individual's emotional engagement, and attachment dynamics.

Likewise, I had the impression that probably Rahmat's and Umar's mothers
had possibly acted not by "fusing the perceptions of their sons with their own

intrapsychic representation of conflictive internalized objects," in Katherine Ewing's terms, but by trying to push (if unconsciously) their sons into a choice that would be more advantageous for themselves within the cultural and social setting wherein they were living (though this must remain a speculation on my part, not being able to interact with Rahmat's or Umar's mothers). To reiterate, I think that in these cases what might be mistaken for intrapsychic enmeshment could perhaps be better interpreted as the demonstration of the profound inroads that cultural schemata and social arrangements make into one's psychic dynamics.[2] The pathological character of what ego-psychologists, and some anthropologists, see in intrapsychic enmeshment, I believe is absent in this kind of culturally induced shaping of individual subjectivity.

Such form of subjectively experiencing one's own representational world as in line with the cultural representational world inherent in their sociocultural environment should not be seen as a pathogenic loss of boundaries in an otherwise well-defined and bounded ego. Indeed it seems that the stress on intrapsychic autonomy or enmeshment should also be considered within the framework of a Western-oriented psychoanalytic thought, which preoccupies itself exceedingly with ego boundaries and defined individuality, following Western cultural schemata that traditionally reward the expression of personal features like autonomy and independence. The analysis of cases of family conflicts, like Umar's and Rahmat's, in the context of non-Western milieus, where cultural idioms of interrelatedness and connectivity may call for different expectations as to how one should behave (and feel) toward oneself and the family members, I believe might be important in order to radically revise the crucial place that in Western psychological thought has been given to concepts like "individual psychic autonomy."

As psychoanalysts like Malcolm Owen Slavin, Daniel Kriegman, and Alan Roland, as well as anthropologists like Charles Lindholm, Robert LeVine, and Suad Joseph, have remarked in the past, it might be more useful (and in line

---

[2] I am not denying here that a pathological form of loss of self-acknowledgment and "self-sensing," with a consequent and persistent conflation on one's own reality with the realities of what is "not-me," may happen in some cases (a "not-me" identification, as posited by Harry Stack Sullivan. Sullivan 1953: 158–164). What I believe, however, is that these cases should probably considered as extreme cases of psychotic or borderline psychotic instances, as Wilfred Bion analyzed in great details, and which should not be taken into consideration for the explanation of psychological dynamics in socially functioning and relatively well-adjusted individuals within a non-Western cultural milieu. With regard to the necessity to differentiate between "pathological" (for lack of a better term) fusions, or mergers, and the functional and adaptive (i.e., "non-pathological") interpersonal encounters of subjectivities and perceptions, the psychoanalyst and developmental psychologist Daniel Stern has written powerful pages (see Stern D. 2001 [1985]: 104–111).

with cross-cultural empirical evidence) to assume that the individual self (or ego) may be in fact constitutionally composed at least in part by the "selves" of other people (i.e., those individuals who have had, and have, the most relevant impact on our emotional, social, and cultural lives), instead of maintaining as a working hypothesis that the self (or ego) is bounded and "isolated." Incidentally, this view may be reconciled also with part of the ego-psychological literature, for example with the model of the representational world that Sandler and Rosenblatt proposed in the 1960s, which I have sketched above.[3]

The context of masculinity: between fantasy and reality

How does masculinity fit into the picture that we have so far painted? It is clear that Rahmat seeks to embody and represent in the most complete manner the ideal image that Pashtun culture has created for a respectable, honorable, and virile man—a hegemonic masculinity. Or, in other words, Rahmat manages to perceive himself in line with the requirements and expectations of his sociocultural milieu, which, as I have noted, does not leave much room for alternative, and competing, "hegemonic *masculinities*." The relationship with his mother that emerged from his narrative helps us understand better what this ideal may imply. While Rahmat's discourse about how "to be good at being a man" touches on the many aspects that we have seen so far both preached and practiced by other informants (ghairat, aggressiveness, restoring a tainted honor, etc.), Rahmat's life events corroborated one more aspect that we have seen already emerging in Umar's narrative: the powerful and influential role of the mother. There is clearly, for Rahmat, no contradiction between his self-representation as a tough, frugal, and "independent" man (in the sense that he does not want to depend on anybody else, and refrains from asking support in times of need, unless in extreme cases), and the fact that he quietly and timidly acquiesced to his mother's request to marry the girl his father wanted him to marry—all the more remarkable because of the ideas about women that he openly expressed with me, as being weaker individuals both physically and psychologically.

---

[3] See, in this regard, the powerful views expressed on the issue of attachment in cross-cultural perspective, and the dynamics of emotion-creation, in a recent collection of essays co-edited by Hiltrud Otto and Heidi Keller (Otto and Keller 2014). The chapter authored by Birgitt Roettger-Roessler is particularly pertinent to the analysis I propose in the section above.

Yet the contradiction seems to be quite obvious to a non-Pashtun, like me, listening to the accounts of Rahmat's life's emotional engagements. And Rahmat is not alone among my informants to have expressed such feelings of admiration and esteem toward their mothers. Both Rohullah and Umar spoke about their mothers as strong-willed persons, to whose decisions they, one way or another, did submit at some crucial point in their adult lives. Baryalay lost his mother too early for him to have been able to construe such type of narrative about her, yet the respect he devotes to his wife (of which I was personally witness multiple times) might be an interesting clue to his potential attitudes toward motherhood. So, if the hardships of motherhood are somehow morally "rewarding" in the eyes of a respectable Pashtun man, where does the apparent cognitive dissonance come from, which allows Rahmat (and many others like him) to think of "average" women as "weaker," almost "inferior" beings, but of his mother as an eminently influent and implicitly powerful person, so much so that her wishes must be fulfilled by him? Is it perhaps the self-serving illusion according to which "my mother is not like any other woman"? This seems hardly to be the case, given that all my informants were very much aware that their mothers, as illiterate and uneducated women, excluded from the public sphere, did not in fact represent any exception to the profile of the average Pashtun woman. A special attachment to one's own mother should probably not constitute a surprise in any cultural environment, including Pashtuns, nor should a culturally shared representation of women as deficient in some way vis-à-vis men. However, in the case of a cultural environment such as Pashtuns, which expresses clearly its heavily androcentric character (even "hypermasculine," from a Western standpoint, with a consistent predominance of a specific type of "hegemonic masculinity"), and the consequently secondary social position held by women, the recurrent "unofficial" recognition of women's personal power and strength, if only in the figure of one's own mother, seems noteworthy. It would obviously take a detailed analysis of several cases of relationship between sons and mothers to start to understand how such phenomena are played out at a personal level, and how much the cultural idiom of masculinity and general social arrangements proper to Pashtun society influence the private negotiations of these intra-family processes.

One general conclusion that we can take out of all this, however, even in the absence of detailed mother-son relations case studies, is that men's subordination to mothers' wishes and expectations is not considered to be a point of weakness, or a "feminizing" aspect of one's personality traits. It rather appears to be a well-entrenched feature of Pashtun masculinity that makes a nar of a simple man (saray).

Rahmat, during our conversations, did not make a secret of those main requirements, moral and ethical, that a nar, in his eyes, should display.

> *Honor [izzat] and land go together. Land is a symbol of your honor. Selling land is a disgrace . . . nobody should sell their land. You see, the more land you have, the more honor . . . land is better than money: money comes and goes, land stays. A big family, with a lot of land, they are respected. Money is not as important . . . only stupid people respect money more than land.*
>
> ANDREA—*I see a lot of people who are very interested in money, and think that respect comes from money. . . .*
>
> RAHMAT—*Yes, there are now, there are these people now around . . . it's because of the war, before things were different . . . now people are scared, they are afraid of what will happen, they changed their way of thinking . . . you don't know if tomorrow you will be dead or alive . . . people want and respect money today because they want things now, right away . . . tomorrow who knows what happens. . . . They think yes, I have this piece of land, and I will keep it, so that people respect me . . . but maybe another war comes, and I die next month . . . so, then? What did the land serve me for? I am dead anyway. . . . But respect should not come only from money . . . many people still respect wisdom. . . .*
>
> ANDREA—*What about you? Do you have land?*
>
> RAHMAT—*Yes, you know about the land we have in the bazaar . . . that's valuable, it's not easy to find land in the bazaar, it's very expensive now . . . that's why the 24 families wanted to take it away from us . . . then we have another big piece of land, about 80 jiribs, outside of the village, toward the mountains . . . that is also valuable, maybe 30,000 dollars per jirib, but it's contested, there is another family that says that a piece of it is theirs. . . . I don't know how it is going to go . . . with my share I could send one of my brothers to live in America or Europe and go to university there. . . .*
>
> ANDREA—*But you told me that only stupid people sell their land, and that it would be a disgrace. . . .*
>
> RAHMAT—*I know what I told you . . . sometimes the situation is so bad that one has no other alternative. . . . I want my brothers to have a good life, to get educated.*
>
> ANDREA—*So, you are going to leave your village too?*
>
> RAHMAT—*No, not leave! Why leave, anyway? I know there are problems, fighting, conflicts, but it's nothing serious . . . you have to be a man, you*

know? A man faces his problems and does not run away, if the problems are within your possibilities . . . here in the district, there are more than 400 Waraki, but only a few of them are ready to confront a problem, like men, when it happens. In Afghanistan, people tread on your rights every day, and the fact that there is no government forces you to defend your rights with your own hands. . . . When I was young, me and my friends would all help each other . . . we would come in support of each other when one of us needed help, we would fight for each other. Now most of them have found good jobs, they have left the village, they have gone to the city, or somewhere else, they have a good life now. They don't call anymore, we don't see each other anymore . . . but I don't like to ask for favors, a man does not beg, I am not a beggar, I don't want to put pressure on people. . . .

ANDREA—*So, you are planning to sell. . . .*

RAHMAT—*Maybe, I don't know, we'll see. . . . I want at least one of my brothers to go away, to go somewhere where he can study, and have a normal life. Maybe Europe. . . . [long pause] certainly not a Muslim country, where there is no respect for people's rights. No, he has to live where people respect him and his rights. . . . A friend of mine from the US, he is now in Costa Rica, he has a business there, he told me that if I can go there we can work together. . . .*

It is remarkable that this last exchange opened with Rahmat saying that there is no honor without land and that selling one's land is a despicable act, but closed with him pondering on selling his own share of land to send his brothers abroad (and himself to Costa Rica). Rahmat did not realize the incongruence when we spoke. His thoughts were flowing, and that's where they took him. From the idealized realm of the cultural idiom and his heartfelt pragmatic and performative adherence to it ("you have to be a man . . . stand up for your rights . . . a man does not beg. . . ."), he shifted to the realm of fantasizing, a purview of wishes that he barely manages to justify through some rhetorical gymnastics ("sometimes the situation is so bad that one has no other alternative. . . ."). After all, it seems, his self-representation is not confined only to self-images and subjective states that are in a direct relation to those cultural idioms about "how to be good at being a man" that he otherwise, and authentically, cherishes dearly. Or is it really so?

He holds a "fantastic" self-image that seems to entail rebellion and liberation from a harsh sociocultural environment and its constraining norms—in Costa Rica, of all places, a country about which he probably knows nothing, but that he construes as a repository of that degree of personal realization and fulfillment that he feels somewhat curtailed at home. Perhaps Costa Rica represents that place where paradoxically he could be, at last, the true Pashtun man that he feels he cannot be any longer in his own Afghanistan, a country plagued by a war that, he said multiple times, changed the minds of its people, betraying his own expectations. If so, it would be the triumph of precisely that self-image of the quintessential Pashtun man, adherent to the cultural idiom of it, that he built for himself over the years and that he feels he cannot fully embody in Afghanistan. The situation, he says, sometimes forces you to come to extreme measures. He would not become like those who sell their land only for money, and self-aggrandizement, in a shameful way—his case is "different." In the end, then, the interdependence between the sociocultural realm and the individual's psychic realm seems inescapable and complete. The psychic processes that make him fantasize of a new beginning far away from home are still fully dependent on, and shaped by, the cultural material to which he was exposed during his own life (the cultural representational world of his social environment, as we have termed it above). Such is the language with which his psychological dynamics speak. Their patterns of functioning need not be peculiar per se, yet the content is culture-specific. As Edward Sapir understood long ago, "the individual in isolation from society is a psychological fiction" (Sapir 2002: 233).

And yet Costa Rica is a fantasy. These diverse subjective states clearly coexist with each other, even compete with each other. These self-images represent the outcome of the interplay between cultural narratives (which he has absorbed and elaborated over time), and the painful real-life situation that he is obliged to confront every day in his rural district. The cultural norms and rules about how "to be good at being a man" give him a moral compass that he has positively endorsed and made his own. The disruption brought about by the conflict inspires in him wishes and strivings that such morality may hardly accommodate.

Again, for Rahmat, like for Umar, Rohullah, and Baryalay, there is multiplicity, shifting subjectivities that still manage to constitute, in spite of inner conflicts and ambiguities, a coherent and psychically functional selfrepresentation. In an influential article, psychologists Hazel Markus and Paula Nurius wrote about the importance of what they called "possible selves": "Possible selves represent individuals' ideas of what they *might* become, what they *would like* to become, and what they *are afraid* of becoming" (Markus and Nurius 1986: 954, emphasis in original). The same applies for those subjective states (or selves, in the authors' parlance) that one endorsed in the past, and/or still endorses in the present. Rahmat was expressing to me the existence, or rather, coexistence, of such "possible selves," in the form of fantasies, and in the language of his own cultural representational world.

Conclusion

The vicissitudes of Rahmat's life shed some light on a different type of personal profile from the ones that we have so far encountered. Rohullah, Umar, and Baryalay all, for different reasons, and in different ways, clashed with and rejected at least part of the cultural representational world they received from their social environment—particularly in the purview of idioms of masculinity. They struggled and adjusted to their social milieus. With Rahmat, we have witnessed a radically different plot. Both during our sessions and in the glimpses of his real life that I could experience personally, Rahmat by contrast appeared to embody perfectly the ideal of the "real" Pashtun man. Those who know him well shared this impression. Rahmat never bragged about this aspect of himself, or tried to show *me* how much he was in line with the "correct" model of Pashtun man. All his accounts about himself and the milieu were presented in a rather unassuming, matter-of-fact fashion. This struck me as peculiar. Although his life trajectory was far from being smooth and even, still I had hardly ever met a man who appeared, and reported himself to be, so well adjusted to, and in tune with, his sociocultural environment—especially when it came to matters of "being good at being a man."

Was it a macroscopic instance of self-deception (as evolutionary psychologists would call it), a case in which the subject is not able to relate certain open symptoms of malaise to the probable underlying, unconscious causes? I was not inclined to give a positive answer, given that Rahmat had showed multiple times to be a very sharp and introspective individual, in spite of his deficiency in formal education. And yet, the fantasy that Rahmat constructed, to leave Afghanistan and go to Costa Rica, constituted in my view a red flag for the existence of a deep sense of frustration and maladjustment, in spite of the apparent positive acceptance of, and adaptation to, his own sociocultural environment, which Rahmat showed publicly. He construed this frustration in terms of an act of betrayal that the environment itself had "carried out" against his dreams and wishes of a culturally appropriate "manly" life at home. It is possible that the identification between the cultural and the personal representational worlds in Rahmat became so strong and "authenticated" by deep emotional engagement to foreclose for him the possibility to remain in his own life at home.

From this standpoint, in comparison with Rohullah, Umar, and Baryalay, Rahmat constitutes a different embodiment of the rejection of a cultural status quo that he feels does not represent his subjectivity any longer. If Rohullah, Umar, and Baryalay rejected altogether that part of the Pashtun cultural idioms of masculinity, and social dynamics, that the decades of war had bequeathed to them as a dire legacy, and became somehow hostile to what they interpreted as a "traditional" Pashtun idea of manliness *tout court*, Rahmat seems entrenched in the cultural idioms of what he feels to be a more "authentic" Pashtun manliness, a more "original" nartob, in the way many of my middle-aged informants recalled it while reminiscing about the pre-conflict years. Rahmat dreams of a village that still works socially and culturally along the lines that he got enculturated into by his father and grandfather, and which he saw operating as such during his childhood years in the family enclosure in Pakistan. Rohullah, Umar, and Baryalay have absorbed and reacted to their cultural milieus in very idiosyncratic ways, and their private, personal negotiation of that cultural representational world is testimony in them to the profound interplay between the cultural and the individual realms. They participate in the societal feedback process, and in the personally induced, almost imperceptible yet constant contribution to sociocultural change through their public behavior and social interaction.

Rahmat does not escape this process, but his contribution probably moves toward a different direction. He has privately discriminated rather clearly between a cultural idiom of masculinity that belongs, in his view, to a "respectable," almost heroic Pashtun past, and one that he considers morally bankrupt. His entrenched rigidity in upholding by example this idealized "respectable" set of values and qualities that supposedly make up a nar contributes just the same to sociocultural change through a feedback effect, yet in a direction that points to the "restoration" of an idealized pre-conflict Pashtun past. Militia commanders, their moral and ethical bankruptcy, the ever-increasing tide of free-wheeling violent behaviors in rural Pashtun areas, have no place in Rahmat's prototypical world, and those around him are aware of it. In spite of his own underlying sense of hopelessness and disillusionment, which leads him to fantasize about a new life in Costa Rica, Rahmat's role as an individual bearer of sociocultural change within his daily life milieu (if imperceptibly and unconsciously) is not forfeited.

6

Between What "Was" and What "Is"

Four Tales of Development and Growth

Introduction

The present chapter will be devoted to four informants whom I have known for a long time but with whom I did not have the regular series of one-on-one interview sessions, as I had with the preceding four informants. Kamran, Inayat, Wahid, Nasim Khan (and his father Niamatullah, to a lesser extent) are among the friends I know best in Afghanistan. I have met them on a regular basis since 2009 (in the case of Nasim and his father), or 2010 (in the case of the other three), and I had the opportunity of knowing many of their (male) relatives, siblings, and friends. Nevertheless, they showed themselves disinclined to embark in the sort of commitment that my "person-centered" project would have required. They simply did not seem interested in subjecting themselves to the admittedly rather intrusive, and possibly anxiety-engendering, practice of sitting down for hours with me talking about their intimate lives, fears, conflicts, and wishes. Yet we spent a good deal of time together, holding our conversations at each other's places, in restaurants, during leisure trips, at weddings, playing soccer, or maybe just while taking a walk together in their respective town or village. From the time spent together I feel that I gained a good insight into their private lives, as well as into their public ones.

Inayat and Kamran come from a small rural village near the border with Pakistan (Sama Khel), while Wahid and Nasim Khan were born and raised in Jalalabad. Though all of them are Pashtun young men, they hail from different sociocultural backgrounds. This puts them in different analytical categories. Claiming that Jalalabad represents a cosmopolitan and "liberal" milieu would be certainly an exaggeration. The social fabric of the city, especially due to the flight of much of the educated and (relatively) secular middle class during the decades of conflict, has become almost undistinguishable from that of a rural district ("Because of the war, the village has conquered

Crafting Masculine Selves. Andrea Chiovenda, Oxford University Press (2020). © Oxford University Press.
DOI: 10.1093/oso/9780190073558.001.0001

the city," one of my informants told me once, with visible regret). Yet the "social ecology" of the city life does impact its inhabitants and creates an environment that in turn attracts a self-selected section of the rural population. In the midst of the high number and high density of city dwellers, between those who permanently live in the city and those who use it only as a temporary hub for working and trading opportunities, one's original rural social networks may loosen their grip. The control villages normally exert on the individual may over time start to fade away. Surrounded by strangers, one can quickly slide into anonymity. While this can bring about feelings of estrangement and alienation, it also takes the individual away from the ubiquitous and inquisitive gaze of the "other"—an "other" who, in the village, knows everything about your life, your family, your lineage's past vicissitudes, and eagerly awaits your next social misstep. Thus, the panopticon to which a Pashtun man is tied disintegrates somewhat in the city. With less social policing by the community, an individual finds he has leeway to enact certain behaviors that in the village context would be frowned upon. Social activities (like schooling for girls or civil society organizations that promote political consciousness or women's civil rights, for example) can be started and supported by the most entrepreneurial members of the urban community. Those who cannot stand the harsh and demanding life of a rural village often find some respite and personal realization by moving to Jalalabad.

At the same time, however, the observer must remain careful not to overestimate the latitude for "deviant" actions among the city dwellers. Whereas the cleavage between rural and urban milieus is certainly present in a province like Nangarhar, the two are closer after almost forty years of war. The complaint "the village has conquered the city" is not an isolated cry, but the confirmed perception of many middle-aged Pashtuns in Jalalabad.

Against this background, Nasim, Wahid, Inayat, and Kamran certainly have a lot in common. They are the representatives of a young and "upcoming" generation of Pashtun young men, who grew up in the context of the post-Taliban Afghanistan, and in the wake of the powerful ideological inputs and influences brought along by the occupation of the country at the hands of Western apparatuses and personnel (both military and civilian). They all show, each in his own idiosyncratic way, how pre-existing cultural narratives are being invested with new meanings, creating new avenues for social, and culturally legitimate, self-fulfillment.

With the possible exception of Nasim Khan, the protagonists of this chapter are much younger than the individuals I presented previously. Even more important, all four of them never left the social environment they were born into. This continuity in personal life experiences and the contiguity of the type of sociocultural milieu in which they were all raised makes for a better understanding of similar patterns in their life trajectory and personal choices. More than just complementing the more detailed profiles I presented in the previous chapters, the discussion of these young men's lives represents a clearer example of how the "new" and the "old" get incorporated into each other, and how the individual manages to shape a different and alternative subjectivity, which transcends both what "was" and what "is."

Inayat and Kamran

When I met Inayat and Kamran for the first time, in the summer of 2010 in their village of Sama Khel, they were just sixteen and seventeen years old, respectively. Neither of them knew for sure his exact age, but, judging from the fact that they were in their tenth and eleventh grade in school, they reckoned they should probably be approaching adulthood (Kamran was slightly older than Inayat). They were among those whom I met visiting the hujra of Hajji Wahidullah, the elder of the Waraki family to which Rahmat also belonged. Inayat and Kamran were the younger cousins of Rahmat, while Rahmat and Kamran are grandchildren of different brothers of Hajji Wahidullah. Inayat is the direct grandson of Hajji Wahidullah himself. Brought to the village by Hajji Zia (another grandchild of one of Hajji Wahidullah's brothers), my wife and I were quickly introduced to Inayat and Kamran. The two were drawn to us because of their ability to speak English. In fact, Kamran even taught English to children in the village who wanted to start with the language, and his classes also attracted students from the neighboring villages. He had taken English courses in Angur Bagh, the district capital, and now had started an independent "practice," which was well rewarded economically. His English was quite good, although it obviously lacked the features granted by practicing with mother tongue speakers, which he had never done. Given the circumstances, his proficiency was rather remarkable. Inayat, on the contrary, was in the beginning stages of learning the language and was at that time one of Kamran's students. This put Inayat in a curious relational position with Kamran. Inayat clearly acted with deference and timidity toward his older cousin (and teacher), but it was Inayat who would hold a higher rank in the family politics when the time comes. He is the direct grandson of

Hajji Wahidullah. As such he is one of the most likely candidates for replacing Hajji Wahidullah as elder of the family, in due course.

I had to wait to understand this, though, until Inayat himself told me one day, without much fanfare, and in a very unassuming way. Kamran was the older cousin of the two, and also the most "aggressive," to whom Inayat naturally deferred. Inayat acted as a very shy and respectful adolescent (toward us as well), with a discretion that was proper of older individuals. The two were obviously two very different characters. Kamran was daring, outspoken, even bragging sometimes, more "attentive" to the shared cultural idioms about the hegemonic masculinity that his milieu upheld. Inayat, on the contrary, was more introverted, soft-spoken, less outwardly sure of himself, and less "performative" in the ways he would be supposed to behave as a future lineage leader. Perhaps part of this might be attributable to the necessity that Kamran possibly felt to "prove" himself to his familial context (in his capacity as a member of a secondary branch of the lineage), while Inayat could feel more "reassured" in the light of his ascriptive position as a direct descendant of the current leading elder. I could not gather enough information about the family environment in which both boys were raised, which surely had a big role in their adolescent development. What was evident, though, was that Inayat had internalized more the values of hierarchy and submission to authority that are integral parts of Pashtun social relationships, while Kamran had "chosen" (unconsciously) to privilege those values of individualism and personal competition that are equally, if contradictorily, a crucial aspect of Pashtun masculine culture.

In the quiet summer of 2010, the seventeen-year-old Kamran was the boy Hajji Wahidullah "assigned" to me, in order to spend time with me, to make sure that I would not lack anything as a guest. He accompanied me for walks in the bazaar of Angur Bagh and introduced me to some of his friends who either worked in shops there or hung out with other friends along the bazaar stalls. He was happy he could practice his English with someone he considered a "native speaker"—I was for sure the closest to a native speaker whom he had had any relationship with. Sometimes Inayat would come along with us, although it took some time to convince him to speak out and not be shy about his less-than-perfect English. While Kamran was happy to show me around the village and the bazaar, as well as to show off to his friends his acquaintance with a foreign person (which clearly scored him a lot of points among his peers), he was equally careful of not exposing himself too much, and kept my presence somehow quiet. He was cognizant enough, in spite of his age, of the way the village worked, and of the hostile attitude that many adults had toward foreigners and outsiders in general. So I was introduced to his friends who worked in bazaar shops, but our conversations took place inside the shops, sometimes even in the back room, where passersby or customers could not see us and ill-meaning eyes would not pick up on my presence. He instructed me to speak to him only in Pashto when we were in the streets of the bazaar, or, better, not to speak at all. Over time, the frequency of our excursions to the bazaar, and around the village, decreased steadily with the increasing deterioration of the security situation in the area. Kamran started avoiding the main bazaar road and entered his friends' shops only from the back doors. By the summer of 2013, both Inayat and Kamran wanted to meet me only in Jalalabad, at my place, where there would be no danger for either of them to be seen in my company. And even in Jalalabad, one thing that Kamran and Inayat always asked of me was not to accompany them up to the taxi stand from which the cars to the village would leave. They wanted to avoid the remote chance that someone from the village, or from Angur Bagh district in general, going back from Jalalabad, could identify them in the company of a stranger, possibly a foreigner. I would usually pay for their taxi ride back and forth from the village, but would walk them only halfway between my home and the taxi stand in the city center.

Kamran was bold. From the first, he was asking me and my wife whether we could buy him and bring him things from abroad. It started with books for learning English ("with a CD for the pronunciation!," he made clear) right the first time we met, and proceeded with a computer ("to show my students some DVDs in good English," he explained), and later in our relationship, even more directly, money ("you should give me some money if I graduate from twelfth grade with good marks!"). In this regard, he definitely strayed from the ideal model that a respectable Pashtun man should embody, as described aptly by his older cousin Rahmat: never ask for anything, never ask for help, do not beg, and try to get by through your own means. It seemed that "begging" was not a problem for Kamran, provided that I did not tell anything about it to anyone ("Don't tell anyone I asked you for these things, make it as if you were giving me a gift"). His boldness was even more evident because, despite his young age, he had an income as an English teacher from students who paid a small "tuition" to attend his class. Nevertheless, Kamran could boast that he earned money for himself. How much of this money he kept I could not ascertain, but I am sure his "craftiness" helped him to keep some of it for himself. I played along and procured him an English textbook, a secondhand computer from Kabul, and later even gave him a monetary gift after he showed me the certification from his high school attesting that he had in fact passed twelfth grade with good marks.

In exchange for these "gifts," I was able to participate in three of his teaching sessions in the bazaar classroom (during which I was put under fire by the students' questions about myself and my work), and, more generally, to keep him wanting to see me on a regular basis. It was clear, in fact, that for Kamran our relationship was based on a premise of reciprocity, which did not disturb me a bit. I could well understand that he had picked up on the common myth, very popular in Afghanistan, according to which every Western foreigner is a rich man. "He is a rich person," Kamran probably reasoned, "and he needs my help. So, I will get something in return for my help. That's only fair, he's got more money than I will ever have, I am not hurting him."

Though I perfectly understood this line of reasoning, and was not surprised by it, at the same time it ran counter to the expected behavior of a respectable Pashtun man, as Rahmat explained. Moreover, I represented the guest, a traditionally sacred figure in Pashtun culture, from whom a host should not ask anything in return for hospitality. Yet, as I would learn during my fieldwork, sometimes, and in someone's view, the foreigner falls outside the purview and the boundaries of appropriate cultural behavior. Bound by strict cultural norms in the relationship with their peers, some of my informants seemed to enjoy the possibility of "violating" the norm with me, particularly when the violation could bring them some sort of benefit—a benefit they could not reap in the standard peer-to-peer relationships. Not that in Pashtun society individuals will be expected to do things only out of the goodness of their hearts, of course. Rewards and compensations are expected, and people exchange favors also in view of some future benefit. Yet this usually happens in a subtle, very diplomatic, and almost unspoken way, far from the brazen fashion in which Kamran asked me to buy him books or computers, not to mention the handing over of money itself.

This aspect of my relationship with my informants took an interesting shape. The existence of, and abidance by, the norms and tenets of good Pashtun etiquette and ethical behavior clearly did not stifle or hinder the existence of wishes, desires, emotions, and fantasies that would be considered socially and culturally inappropriate. Kamran would have never asked for gifts, the way he did with me, of an adult Pashtun man. In the presence of his elders he was always very deferential, obedient, and respectful, never speaking if not addressed to. He would never ask his older cousin Rahmat, or the brother of his father, or a friend of his father, for a computer despite the fact that they probably could have afforded it. He could do so with me because, I believe, he deemed that I was unable to judge his actions (and so himself) in a culturally competent way. In Pierre Bourdieu's terms, he thought I was not "fluent" in the cultural language that gave or denied him social capital. From a certain standpoint, he was taking advantage of what he supposed to be my ignorance. What most interests me in this case is the realization that his rigorous training in proper etiquette and ethics did not prevent him from "feeling" in a certain way, from having certain wishes that he knew were socially inappropriate, and which he could satisfy with me—asking me for gifts and money, displaying a daring and "impudent" attitude that perhaps will serve him later in life as well among his peers, if he ever reaches a distinguished enough position in the lineage.

Kamran was a "naturally" entrepreneurial boy. The effort he put into learning English, as a way out from the village in a near future, and as a means for income in the present, was remarkable and successful. Inayat confessed to me that, among their friends, Kamran acted in a domineering and aggressive way. He was "stronger" than the others. In this sense he was fulfilling the other half of Pashtuns' classical equation for proper masculinity, at least in a rural context: boldness, courage, and daring (the first half being respect for authority and capacity for self-effacement). As noted, I always had the impression that he felt he had to overdo things in order to gain a social position of repute that by ascription would go to somebody else (in this case, probably Inayat, as the grandson of Hajji Wahidullah). When asked directly about certain cultural precepts and social customs, his opinion was often very strong, as if looking to assert a strong and uncompromising ideological or identity stance. Once, in 2010, we were standing in the area outside the indoor village hujra, against a hot August sunset, and I asked him whether there were any of the women living in Sama Khel who had a job outside the village. He replied:

No, of course not, women should not work! A woman who should go to an office to work, or somewhere else to work alone, away from the village, she should be shot!

ANDREA—*Shot? You mean that you would kill her?*

KAMRAN—*Yes, that's the only thing you can do in this case . . . she dishonored the family, she had relations with other men . . . that's what should happen, she should be shot!*

ANDREA—*What do your father and uncles think about this? Would they agree with your opinion?*

KAMRAN—*Well, I don't know, sometimes they tell me that I should not say certain things, that I don't know what I am talking about . . . I think they would not support me on this. . . .*

ANDREA—*So, where did you take these ideas from?*

KAMRAN—*But this is pukhto! This is Islam! A woman should stay at home, and not have any relationship with any non-muharram* [non-close relatives]. *There are many people who say that the punishment for a woman who dishonors her family can only be to kill her.*

ANDREA—*What happens if her family members don't kill her?*

KAMRAN—*Then they are beghairata* [without honor, unmanly], *they are all beghairata, especially the husband . . . and the woman is* behaya, *she is also dishonored* [behaya means literally impudent, without the proper moral education, which may apply also to men].

This was Kamran at seventeen years of age—a smart and quick-thinking (and quick-acting) boy with a lot of ambition and a lot he wanted to demonstrate to his peers and social milieu in terms of cultural appropriateness—even against, apparently, his own elders' best judgment. The bold and uncompromising profile that he chose for his public self, possibly after the example of the more radical figures among his acquaintances, is in contrast with the example that Hajji Wahidullah sets within the family. A soft-spoken, quiet, reflective, and balanced character, the main elder of the Waraki lineage clearly incarnates an "old guard" within Pashtun society, to whom some young, upcoming men like Rahmat still look for inspiration. And it should not come as a surprise, although it might seem counterintuitive. It has happened multiple times to me to encounter sixty-, seventy-, and even eighty-year-old lineage leaders from troubled rural areas of Paktia and Nangarhar provinces displaying an open-mindedness, sophistication, and bluntness in their social and cultural ideas that many of the following generations, raised during the conflicts, failed to show. "Women are those with whom our children spend more time. An uneducated woman will raise only ignorant and foolish children," I was told by an illiterate seventy-five-year-old man who hailed from a faraway district of Ghazni province, at that time completely under the control of so-called Taliban insurgents, who forbade girls from attending any kind of schooling activities. "I sent my daughters to school," he continued, "then came the war, and the Taliban . . . my granddaughters did not go to school . . . this is wrong in Islam," he added.

The generation of Kamran and Inayat (but not necessarily each, as individuals) was exposed to a shifting set of cultural and religious idioms about the proper policing of social arrangements, and the display of appropriate masculine personality traits, that had its roots in the deviant violent behaviors institutionalized by the war, together with the irruption of radical religious ideologies that came along with it from outside Afghanistan. The competition between these two threads in current Pashtun society is evident today in the different profiles we can find in the very same family, such as Rahmat's and Kamran's. Although Kamran was at that time too young to have done more than express in words his desire to appear publicly in a certain way, still his tirades are somewhat indicative of an effort at self-representation that owes more to a recent, war-impacted moral landscape than to a more "traditional" Pashtun cultural milieu (as evidenced by my older informants, and those adult ones who identify more strongly with them, like Rahmat).

Over the years I have seen Kamran change, as well as Inayat. The shift from late adolescence to early adulthood was noticeable in both, particularly in Inayat. In the summer of 2013, Kamran and I had one of our last meetings, in Jalalabad, over a good Friday lunch in a restaurant downtown, after Kamran had attended midday prayers in a mosque near my house.

ANDREA—*Do you remember what you told me one of the first times we met, three years ago, in the dera outside Hajji Wahidullah's hujra? You said that a woman who works outside of the house should be shot, that women should not work away from their husbands, remember?*

KAMRAN—*Yes, I remember . . . why?*

ANDREA—*Were you just telling me the official version of the story, how things should work, or did you really believe that women who work should be shot?*

KAMRAN—*It's my conviction, I would not respect a woman who works outside of her home, without her husband. . . .*

ANDREA—*Are there no women in Angur Bagh who do that?*

KAMRAN—*In Angur Bagh there are some, but they work in Jalalabad, where nobody sees them . . . in Sama Khel no woman works, they would be behaya. . . .*

ANDREA—*Are the women of Sama Khel literate [baasawada]? Did they go to school?*

KAMRAN—*No, none of the women in Sama Khel went to school. . . . In Angur Bagh there are some who went to school up to fourth or fifth grade, and if there are female teachers available they can go to school up to twelfth grade, but schools are too far from Sama Khel, it would not be appropriate for girls to go to school alone. . . .*

ANDREA—*What about you? Are you ok in the village?*

KAMRAN—*Well, so far I am ok . . . you know, I just finished school, I teach English, I spend time with my friends . . . but after this, when I won't go to school anymore, when maybe I will not want to teach English anymore, then . . . then life in the village will become hard. . . . I mean, what will I do? I will have to spend all my time in the hujra, in the village, I will hate it, I will get bored. . . .*

ANDREA—*Your family has land: could you not start farming the land of your family?*

KAMRAN—*No, our land is far from the village, I could not do that every day . . . my father and uncles don't farm the land . . . no, I want to go to university, maybe in Pakistan, where education is better and cheaper. . . . I want to leave the village. . . .*

ANDREA—*But you know that in Pakistan life is different from what it is in the village . . . you know that in Peshawar women can work outside of their home, and can have friendly relationships with non-muharram as well . . . you know that in Peshawar University there are many girls as well who go to class. . . . They have a different mindset [nazariat] in Peshawar, you know that, right?*

KAMRAN—*Yes, I know, I know Peshawar . . . but it's ok, I would still judge them badly, and I would still want my wife to behave like she should behave in the village, but I am tolerant of other people's different behaviors, I understand that they have another mindset, I don't have a problem with that. . . .*

So, three years after our first encounter, now in his early twenties, Kamran still professes to be in line with the moral and ethical tenets that his cultural milieu passed on to him. He believes in what the cultural idiom (schema) proposes for a socially appropriate life, he says. The self-image he holds of himself is still firmly rooted in his cultural background—in the cultural representational world that exists around him in the village. Yet, something seems markedly different from the stance that Rahmat, for example, displayed with me. Kamran does not like the village life per se, he does not feel that strong emotional bond that, for example, gathering in the hujra together with the other male family members creates for Rahmat. Likewise, land, if not irrelevant in Kamran's life's equation, certainly does not represent either that pillar for one's own culturally appropriate self-image as respectable Pashtun man as it does in Rahmat's case. Rahmat came back to the village from Pakistan, remained there, and considers leaving the village as a last-resort option only, in case of overwhelming circumstances. He believes that all those who leave their ancestral place because of other than overwhelming circumstances are morally ambiguous individuals, not to be trusted. Kamran wants to leave the village and go to Pakistan, and build a different life for himself. Kamran, as we will see even more clearly with Inayat, and as we have somehow seen also with my previous informants, seems to be representative of a new generation of Pashtun men from rural areas, who, albeit still firmly rooted in their own cultural milieu, feel the latter to be somewhat too narrow, too limiting.

In Umar and Rohullah, certainly the social and cultural inputs received from experiencing different lifestyles and mindsets throughout their lives had a role in such development. Kamran, however, as Inayat, knew only his own village and, only cursorily, Jalalabad and Peshawar. He never crossed his provincial borders toward Kabul, or any other Afghan city, for that matter. And yet, in this boy, mainly grown up after September 11, 2001, something of the sweeping changes that Afghanistan underwent since 9/11 has apparently worked through and left its mark. The "traditional" background that his family maintained and perpetuated, particularly through the figure of its main elder Hajji Wahidullah, probably shielded him, as well as Inayat, from the most noticeable aspects of that shift in moral values and ethical behaviors toward increased violence and abusiveness that the thirty-five years of war effected more clearly in other Pashtun environments and social realities. An outsider, however, as I was, would notice the slight contradiction, or, rather, apparent naiveté, that Kamran showed in claiming to be a tolerant person, while at the same time admitting to despising just the same those women (and men) that in Peshawar he would coexist with, who should not follow his ideal model for a socially appropriate Pashtun comportment (which he still derives straight from his village enculturation process and cultural representational world).

Yet I take this as a clue to the quickly developing inner world of Kamran's, a young man, barely in his early twenties, who feels the pull toward a radical change from his family's traditional life trajectory and locale (he would be the first one in the lineage to attend university), acknowledges the perils that this might entail for his self-image of a respectable rural Pashtun man, and yet is convinced that he will be able to maintain a steadfast adherence to the kind of person he is now—his current self-image, culturally appropriate to the village life. I consider Kamran's most recent personal developments, at the time of my departure from Afghanistan (mid-2013), as an interesting snapshot of a potentially turbulent dynamic in its first stages. Barely out of adolescence, Kamran is quickly changing, yet he consciously wishes he could remain the same individual in a different environment and context nonetheless (which is possibly one of the most intractable conundrums that late adolescence may present cross-culturally).

When I knew Inayat, in mid-2010, he was everywhere where his older cousin, Kamran, was. By the time I was about to leave my fieldsite, in mid-2013, Inayat had left the village, moved to Kabul, found a job, finished high school in Kabul while working, and was on the verge of applying for evening classes in business administration at one of the many private universities of the Afghan capital (thanks to a scholarship his Turkish-Afghan employer provided): quite a leap for the shy sixteen-year-old boy I first met in 2010. Inayat never displayed the boldness and self-confidence that Kamran sported publicly. He seldom spoke spontaneously and mainly observed the lengthy discussions between me and Kamran. Sometimes, when the topics became more facetious, or when I addressed him directly, he showed himself more talkative. Unlike Kamran, he never approached me for gifts, or with brazen requests ("When you go back to the US, will you pay for my tuition in college in Pakistan?" was Kamran's final demand before I left Afghanistan). He showed himself as dignified and modest as his cousin Rahmat and his grandfather Hajji Wahidullah, the lineage leader, showed themselves to me. In this regard, he seemed to have internalized appropriately the social meaning of his position within the lineage, as one of the probable successors of Hajji Wahidullah in the leadership of the family—never beg, never complain, be dignified, always try to get by with your own means.

In 2012, when I came back to Afghanistan for my last, long-term research stint in the country, I found he had relocated in Kabul, after finding a job as an accountant for a construction company led by a wealthy Afghan expatriate in Turkey. I visited him at his workplace, where he showed me his compound, and proudly introduced me to the site leader—the son of the company owner. He worked de facto without pay, but his employer guaranteed him accommodation in a shared apartment, three meals per day, and a small per diem for daily expenses. The employer also promised him to pay for college classes if he managed to graduate from high school—a promise that he later kept.

I found Inayat happy to be on his own, as an adult before his real adulthood started. I also visited his apartment, which he shared with six more people working for the same company, in a newly built high rise in the northern outskirts of Kabul. The living conditions appeared crowded to a Western eye, with three and four people crammed in two small bedrooms, sleeping on mattresses lain down on the floor. Yet they were not dissimilar from the communal style of living, and sleeping, that I had myself to experience many times when hosted in a village hujra, where relatives and friends enjoyed each other's company overnight (in my own apartment in Jalalabad, I too indeed slept every night on the floor over a *toshak*, a thick mat for sitting down, usually while drinking tea and entertaining guests). It was precisely this that made Rahmat's day a happy one—the proximity of one's own family members and closest friends in the hujra.

Inayat had no complaints about his accommodation. He and his colleagues had electricity (though not twenty-four hours a day, but certainly better than no electricity at all, as in Sama Khel), and running water in the bathroom and kitchen (also completely absent in the village). The only reservations Inayat had about his accommodation did not stem from logistics per se, but from the social aspect of it. "I mean, they seem normal people, they are ok ... but I don't know them, I sleep in the same room with strangers, it is weird for me, I don't know what to expect, I am alone here," he said to me when I visited the apartment. His reservations replicated basically those that his older cousin Rahmat expressed about the period in which he worked on his own in Jalalabad and Kabul, and which was the main reason that he left those positions and came back to the village. The issue was not the lack of any privacy or being constantly crammed with other three or four people in one room. It was *who* those people were that constituted an issue. It's not even that he had something bad to say about any of them. It's just that they were not family or close friends. The feeling of familial connectivity that anthropologist Suad Joseph elucidated so well through her ethnographic material, this feeling of one's own privacy being created and inhabited also through participating in the privacy of other people (what Katherine Ewing called "interpersonal engagement," not to be confused with "intrapsychic enmeshment"): this was what Inayat was missing, something that he had been strongly socialized into and that pervaded his self-perception, as well as the inner emotional framework that he developed over the years.

From his situation I gained a perception of both his "individualized" self (a late-teenaged boy who is "individuated" enough to find the strength to choose to leave home, live alone away from it, be left to his own devices through work and school, in order to chase his own dreams in life), and his "interconnected" self (the village kind of person who is ok with sleeping every night with four people in a small room, but wishes they were his relatives). The unease and anxiety that he perceives at his logistical situation does not stem from any "intrapsychic enmeshment," I believe, which he would have acquired from a peculiar socially interconnected village upbringing (as it might come easy to conclude), but was rather the symptom of his interpersonal engagement, into which he had been socialized within his familial context.

After my visit to his workplace, and his apartment, he visited me one day in my house in Jalalabad. He had come back to the village from Kabul for a few days, in order to see his family and friends. He took a taxi from Sama Khel to Jalalabad to meet me. I was quite sure he had something he wanted to ask of me, something I might do for him, but I was mistaken. We only had a good talk and some tea.

INAYAT—*I am staying for a few days at the village. You know, I wanted to see my parents, relatives, and friends. . . . I miss them in Kabul, I am alone there. . . .*

ANDREA—*Is your father happy that you went to Kabul?*

INAYAT—*Yes, he is . . . you know, a couple of days ago he told me, "we spent all our lives here* [in the village], *and look what we have gained: nothing. You have to stay there* [in Kabul]*." He is right, you know. . . . Some of his brothers went to Kabul when there was Zahir Shah* [in the 1960s, probably] *. . . they went to school, one of them became an officer in the army, they had a good life . . . still now they have a better life than us . . . look at us. . . .*

ANDREA—*You did not like the life in the village? . . .*

INAYAT—*No, I don't want to live in the village, it's a difficult life, it's all about fighting* [lanjay], *having enmities* [dukhmani] *with other families . . . you always have to live on the watch, controlling everything that is happening around you. . . . I don't like that . . . two years ago* [2011], *for example, when you were in America, we had a fight with another family near Gulahi* [a close-by village, now a center for anti-government insurgency], *I did not tell you about this. Another family, the Sali Khel, took a piece of land that belonged to us . . . you know how these things work here, we talked about it*

... they just go and take your land ... anyway, Hajji Wahidullah warned them not to farm our land, and many from my family went there to protect the land from the Sali Khel. I was told to stay home with the women, I was too young to fight. The Sali Khel tried to go back to the piece of land anyway, and there was a firefight ... four of the Sali Khel were wounded and had to go to Jalalabad, to the hospital. So, after that, Hajji Wahidullah asked for a Jirga, and he offered to go nanawatay [the public ritual through which one asks for forgiveness, after which all fighting has to cease definitively], *so the thing now is finished. ...*

ANDREA—*Wait, does not nanawatay bring dishonor to the family which does it?*

INAYAT—*No, not always, it depends ... you see, Hajji Wahidullah was smart, he did not want a feud* [dukhmani] *to start, you know, when people go on with revenge forever ... he wanted the problem to end ... and we had won, we had kept the land ... in this case he asked for nanawatay, and after nanawatay the Sali Khel could not continue the fight to get revenge for their four members wounded by us ... so, the Sali Khel now are beghairata two times, once for having lost the fight, and twice for not being able to take revenge against us ... and we kept the land ... this is how you do things in the village. ...*

ANDREA—*But you did not go to the fight. ...*

INAYAT—*No, I did not. ...*

ANDREA—*Were you upset that you could not participate?*

INAYAT—*No, actually I was happy, I did not want to go. ... I don't like to fight ... but if they had asked me, well, then I would have had to go ... you don't have a choice in those cases ... that's also why I don't like life in the village ... sometimes you have to do things you don't like ... bad things ... there is always someone who is against you, always someone you have to defend yourself from ... it's not an easy life. ...*

Inayat's family has land but do not farm it directly. Farming is considered a demeaning occupation, which is usually delegated to landless Pashtun peasants. Most of the Waraki family members are now jobless, like many other families' members, because of the depressed economic situation of the area. They once held jobs in shopkeeping, transportation, and construction, but now economic transactions are slow and there is no way of working for a living. Even Rahmat lost his job at the US Army base. Yet the Waraki have land and its revenues keep the family going. They are not the only family in this situation. Certainly, this unwelcome state of idleness does not help maintain a peaceful coexistence in the community. Scarcity of resources coupled with time to kill makes for an explosive combination. Hajji Wahidullah's lineage and he himself did not look like the ones who would recklessly take advantage of such a situation, but others might. Indeed, the long-term consequences of the "state of exception" (now virtually permanent) created by three decades of conflict have taken a toll on the moral and ethical standards that now many Pashtuns have accepted as appropriate for a respectable public display and understanding of masculinity. During the period in which I have been acquainted with Hajji Wahidullah's family, three inter-family violent disputes over their property were brought to my attention. I have good reason to believe that those were probably not the only ones.

Amid all such social turmoil, Inayat does not feel comfortable with this state of affairs. He decided to leave. His father, apparently out of concern for his son's personal achievements in life (including economic ones), approved of his decision to leave the village and seek better fortune in Kabul. His father's blessing, a prerequisite necessary for the validity and feasibility of any young Pashtun man's decision, likely emboldened him and made him more resolute. Inayat does not feel cut out for the kind of life that he would have in the village. A particular aspect of his temperament might have a role in his feelings. The many external influences that, if with difficulty, reached his faraway village in a tumultuous post-9/11 Afghanistan must have certainly had a role in his rejection of the village lifestyle that his cousin Rahmat so wholeheartedly embraced. Sure enough, as in the case of Kamran, Inayat knew only his village, Jalalabad, and Peshawar before being catapulted in the cosmopolitanism of Kabul. He was not even able to speak Persian, which is the lingua franca of the Afghan capital, and without which he would not be able to socially function there. He told me life is not easy in Kabul for him. His colleagues at work often tease him for his peasant-like appearance, language, and "etiquette." In their eyes, he regretted to say, he is little more than a "savage" Pashtun. Indeed, "traditional" Pashtun values and ethics, especially in the realm of masculine virtues, are still very much represented and upheld within his family—not only by the "old guard" of Hajji Wahidullah but also by middle-aged individuals like his father and some of his older cousins like Rahmat. Inayat was taught the rules of the game and knows how to play it. Yet he has also come into contact with alternative ways of expression and self-representation and different ways of shaping his own self-image. The cultural representational world of his Pashtun milieu is being infiltrated by new elements (like television, outside acquaintances, the Internet, Facebook, the observation of the sociopolitical situation unfolding in his country), after which he might choose to reshape his own self-image. Although he is little more than an adolescent, his own self-representational world has vast latitude for expansion and a strong capacity for incorporating new inspirations. It possesses perhaps more elasticity and plasticity than Rahmat's, and he may play a role in the development of a shifting Pashtun cultural representational world that will hold a meaning also for those who were left behind in the village.

Nasim Khan (and Niamatullah)

Nasim Khan is the only child of Niamatullah, a man in his mid-fifties. Both Nasim and Niamatullah were among the first people I knew in Jalalabad when I first arrived in the city in 2009. We remained good friends until the end of my fieldwork in 2013. Born in Kama, a rich rural district adjoining Jalalabad, into a local family with long-standing ties to the area, Niamatullah relocated soon to the city together with his six brothers, father, and mother. Nasim was born in Jalalabad around 1985. The family retained all the fertile land they owned in Kama and apparently is still harmoniously handling its management and the revenues that come from it. Because Niamatullah's father is still alive, the property has not yet been divided among his seven sons, which will happen probably only after his death. This is usually a delicate moment when a Pashtun families experiences frictions and enmities between brothers. According to Nasim, currently things seem to go well among the seven brothers.

Niamatullah finished high school in Jalalabad and thereafter spent two years in college in Tashkent, then part of the Soviet Union (now Uzbekistan), studying for a two-year degree in political science. He then worked for many years for KhAD, the KGB-like agency for internal security that the Afghan state organized during the years of the Communist regime. He lost his job after the demise of the last Communist president of Afghanistan, Mohammed Najibullah, in 1992 and thereafter spent almost three years in a prison run by the anti-Soviet mujaheddins. Niamatullah himself always told me he worked as a shopkeeper in Jalalabad during the Communist regime. Nasim, however, confided in private that his father was indeed a KhAD officer. This was a well-known fact to the family members and the reason for his imprisonment after the fall of Najibullah. Niamatullah only started working as a shopkeeper after his jail time was over. Nasim recalled almost jokingly the times when, during the struggle between mujaheddin and Afghan state, his father would disappear from the house for weeks on end. Then one day he would appear again on their front door with a long beard and scruffy hair, all tired and wearing worn out military fatigue clothes, to the amusement and surprise of a young Nasim who realized only years later what it all was about. Niamatullah, at the moment of my fieldwork, was heading, as he had been doing for several years, the Nangarhar province branch of a civil society organization. It promoted political and civic awareness among both men and women in Jalalabad and some rural districts of Nangarhar province. The organization was headquartered in Kabul and funded by international donors. Niamatullah managed to give temporary employment to many men and women who were recruited to implement the various projects for which the organization found money. Through his position and latitude for decision-making, Niamatullah garnered a stable following for himself among many educated individuals in Jalalabad who were desperate for any employment possibility. Jobs (even temporary ones) had already become a very scarce and valuable resource. Niamatullah's family itself could rely on the revenues of the roughly 65 jiribs of highly fertile land (about 26 acres) they owned in the Kama district.

Nasim's memories of the conflict are dramatic and not dissimilar from Umar's, who also lived in Jalalabad at the time. Until 1992, when Najibullah's government collapsed, they suffered the brunt of the fighting in Jalalabad between mujaheddin and state forces. Like Umar, Nasim's family some-times lived for days inside an underground bunker they had built under-neath their home to take shelter from the continuous shelling of the city. The presence of his father was intermittent due to his job. After his father's imprisonment sometime in 1992, Nasim continued to live in Jalalabad with his grandfather (Niamatullah's father), who took charge of Nasim and his mother. Niamatullah had three brothers already living abroad (in Russia and Germany), and three more living in Jalalabad. Niamatullah, the third son, was living together with his father, which made the rearrangement of the fa-milial situation less traumatic. Jalalabad suffered less heavily during the civil war (1992–1996) than during the anti-Soviet struggle. After the Taliban con-solidated their power in eastern Afghanistan in 1996, Nasim managed to at-tend classes in the state-owned high school of the city. It was reshaped by the new regime but still working. By the time I first met Nasim in 2009, he was attending classes at the faculty of law at Nangarhar University in Jalalabad, but his future expectations seemed to him already bleak. In 2009, over a lunch of fried fish at an outdoor restaurant for families in the outskirts of Jalalabad, he once told me he had no hopes for the future of his country. The incessant wars had taken too heavy a toll on the social fabric of Afghanistan. "They will start fighting again as soon as the foreign soldiers leave," he said in good English. "All the good people of Afghanistan have left during the last thirty years, only the criminals and warlords have remained." The very fact that he and his family never left Afghanistan undermined his claim. People throughout had made the difficult decision to remain over the years, so his view on the future might have been too grim. Even after that, more recently, his personal situation improved considerably; nevertheless he never abandoned this somber, reflective way of looking at things. Given such state of mind, his steadfast resolution never to leave the country, even when he would have had the chance to do so, speaks to a family inclination toward "stoicism" that he clearly inherited from his father and grandfather. In Nasim, as in his father Niamatullah, resilience is strangely coupled with resignation to a seemingly punishing but inevitable condition. To my inquiry as to why he would not consider college programs abroad, like those in India, he replied that he would not leave the land and the heritage that his family possessed and eventually would bequeath to him. In the summer of 2010, we had the following exchange:

NASIM—*Pride is what we have here among Pashtuns. It is what makes us Pashtuns.*

ANDREA—*Do you mean ghairat?*

NASIM—*Yes, ghairat, ghairat and* nang [he is referring to nang and ghairat as synonymous here] *is what we have here. You don't see it openly, in the people, like a piece of clothing that they wear. You see it when they have to defend their rights, their property. Other people who live around here don't have ghairat, like us. . . . Panjabis, Bengalis, Pakistanis, they don't really have ghairat. They tolerate more things than we do, when it comes to their rights. They say, "OK, it's not important, who cares." . . . that's not ghairat. They don't care about land as we do. Land is what makes you respectable, without land you will be considered weak. That's how it works here.*

ANDREA—*It doesn't change?*

NASIM—*No, of course it changes, it's always changing. Look at the situation here. Certain things that should be strong, like respect for your elders and for your parents, sacrificing for your friends, not acting like a zurawar, there are not so many people anymore who do these things now . . . maybe in the villages it is more like this, I don't know. . . .*

ANDREA—*Is it different in the city?*

NASIM—*Yes, of course it is different. Life in the city is different, but it is also because of the war . . . people have mental problems, many people do not behave normally anymore because of the war. There is much more violence, there is much more aggressiveness when it comes to protect your rights. People fight a lot now, it's become like a normal thing. I think it is because of the war, because of all the violence that there has been for a long time here. . . .* [long pause] *If there will be peace things will change. . . .*

ANDREA—*What do you mean?*

NASIM—*Yes, with peace things will change, people will change their behavior, there will be less violence, less aggressiveness. . . . You see, now if you do not respond to an insult, if you do not react with violence to someone who has offended you, in the city they probably will say, "Oh, he is a good man, he is an educated person, he does not want problems for something like that." But if you do the same in a village, they will surely call you beghairata. You see, things change all the time, if a better life will come for those in the villages, like it has happened to us in Jalalabad a little bit, things will change also in the villages. I have no problems with that . . . everything changes, it's life . . . people see that there are other ways of living, they watch TV, now they go on Internet, I have no problems with that, that's life, that's how the world works. Educated people also in terrible places like Khogiany* [a troubled, volatile, and very "conservative" district of Nangarhar province] *think the same way. You have to give people peace . . . things will change.*

This conversation struck me because it opened with rather conventional statements on cultural idioms of masculinity and how Pashtun society was supposed to work in Nasim's opinion (which I do not doubt he genuinely believed). It closed with an increasing "coming out" of Nasim's more private views on the actual situation and possible future developments that somewhat contradicted the "official" version he had given me just a few moments before. Ghairat, nang, pride, defense of one's legitimate rights are all aspects of being Pashtun that Nasim claims as his own. In line with the shared, accepted view of Pashtun masculine ethos and social order as a whole, Nasim has always portrayed himself as a "believer" in this form of cultural idioms. Yet he seems equally aware and accepting of the reality that society is in a constant state of change, and that it is pointless to try to stop the tide. More important, he views such change not only as an inevitable "natural" occurrence but also as a welcome development within a historical conjuncture that, in his opinion, has gathered more negative than positive momentum in the past decades. Nasim knows that those same "natural" attributes of Pashtuns like ghairat and nang will undergo radical rethinking in a society that will be at peace and subject to the inputs of external influences. TV, radio, and the Internet will change Pashtuns' view of the world, and their society, but this does not appear to threaten Nasim's own sense of personal value, cultural identity, and social positioning.

This might have to do partly with the (non-)religious background in which Nasim grew up. His father Niamatullah is certainly not an overtly religious person. During the many afternoons we spent together at my house enjoying conversation with some of his friends, or with him alone, he never paused to pray even when all others did so. In fact I never saw Niamatullah pray at all, in all the years I have known him. We never discussed religion together and he kept well away from the subject. During my fieldwork I have seen a few people who I knew did not care very much for religious matters but nevertheless comply with social norms and publicly display their piety by going to mosque at least once a day, and performing the five standard prayers when in the presence of their peers. In rural areas of Pashtun Afghanistan (and the city of Jalalabad maintains still a strong rural character), the public performance of religious appropriateness is crucial for social validation and acceptance. The vast majority of the people I met would perform their five daily prayers, no matter where they were or what they were doing, and no matter what they thought about religion in general. I indeed once sat in a room in Jalalabad, talking to an acquaintance who was sipping wine while chatting with me, when suddenly he remembered it was time for the first afternoon prayer. He quickly left the room, went to the nearby mosque alongside his neighbors, and then came back to the room to continue talking to me sipping wine. He had to maintain his public image of a good Muslim. Against such background of intense concern for social validation and religious appropriateness, Niamatullah's behavior always stood out as peculiar. He simply seemed not to care. His own friends, when talking to me away from him, told me they believed he was a "bad Muslim." Yet they continued to hover around him (particularly those without a job), likely because of his wealth and political connections. I always figured it made sense that the Communist-trained and -educated former KhAD officer would not pay much attention to religious etiquette and would defiantly display his convictions publicly. He could afford to do so, though, coming as he did from a big land-owning family. Others may not have been able to afford the same luxury—like, for example, those who remained around him hoping for a position or a temporary job in his civil society organization.

Nasim seemed to have followed in his father's steps. I have never seen him pray, either publicly at a mosque or with friends, or in private with me. Like his father, he never discussed religious matters with me, and particularly refrained from asking me anything about my own faith, or the reasons why I would not become a Muslim, as many others did. The contiguity and personal identification that Nasim felt with the cultural representational world of his sociocultural environment (its cultural idioms, particularly with regard to individual masculine attributes) became inflected in a peculiar way by their relation to sentiments toward religion. I felt that the quasi-"secular" (in Western terms) fashion in which Nasim looked at, and interacted with, his sociocultural milieu made his Pashtun-ness more amenable to absorb and tolerate change, with the inevitable sense of uncertainty that accompanies it. Apparently disentangled from the traditional identity between a quintessential Muslim-ness and Pashtun-ness, the cultural idioms with which Nasim claimed to strongly identify seemed to acquire a degree of malleability, of elasticity, that allowed Nasim to reconcile his self-image as a "good Pashtun" with the challenges and insecurities stemming from the quickly metamorphosing social and political situation of Afghanistan.

Indeed, Nasim has no problems in asserting his Pashtun-ness, while at the same time welcoming what he sees to be the inevitable changes that the future will bring to his social milieu. If anything, he himself longs for something different, a different way of "being Pashtun" that he would embrace as a necessary adjustment to a changing global environment. In this regard, in late 2011 we had another interesting conversation. He had just resigned from his job with a Bengali NGO, which he liked very much, because of death threats from a self-defined Taliban-affiliated group. The NGO was asking for a 20 percent yearly interest rate on microfinance loans that it provided to poor families in the province. Interest is forbidden by shari'a law, and the NGO's Afghan employees were all threatened that unless they left their jobs they would be killed. Ironically, the person who called him twice on the phone to threaten him was a classmate of Nasim's from high school, a young man he knew well and whom he did not know had started a militant activity. He reasoned with his old classmate, but after the second call and becoming aware that many other employees had been threatened in the same way by different interlocutors, he decided to quit his position alongside everyone else. He was therefore extremely frustrated when we had the following conversation in Jalalabad:

> *I feel like I have been unhappy all my life here, since I was little . . . the years in the bunker, under the bombs, my father in jail, the civil war . . . thank God we at least kept the land in Kama . . . you know, that's important to us, my grandfather did a good job to keep it with us . . . we never left this place, also when everybody else was escaping to Pakistan . . . we stayed here . . . it was not easy, but we had too much to lose . . . the land, you know . . . you can't lose the land . . . now we still have it . . . that makes me proud.*
>
> ANDREA—*But you still feel unhappy. . . .*
>
> NASIM—*I don't know, it's a feeling . . . it's like I wish I had something different sometimes from my life . . . for instance, my wife: I would like to have a more open relationship with her, you know. . . . I mean, I would like to go out with her, to take her to restaurants, to walk around with her, to do something together, also with our daughter . . . but I can't, she has to stay home. There is not much I can do . . . if I took her out people here would give me* peghor [they would criticize him on moral grounds, accusing him of being beghairata], *they would start gossiping, they would talk badly of my family and my father. . . . I don't want that. . . . It makes me feel like a prisoner, sometimes very angry. . . .*
>
> ANDREA—*But this is how Pashtuns do things, right? This is Pashtun custom. . . .*
>
> NASIM—*Yes, I know, but if people were a little more educated, less ignorant, things would change. . . . I am a Pashtun, and I am proud of it, but I do not think that I become a bad Pashtun if I take my wife with me to the park . . . this is stupid, this is ignorant . . . things can change, you know. . . . I feel more free in Kabul . . . when we go to Kabul, me and my wife, I can go around with her, people don't always stare at you. . . . I feel free. . . .*

To be sure, his recent disappointment at having to leave his job because of religio-political radicalism plunged him in a condition of despair that contributed to make all his life appear "unhappy," even against evidence to the contrary. His wife and daughter, for example, were certainly a source of support and joy for him, as was having overcome the worst of the conflicts with his family's properties intact. Nevertheless, our conversation pointed again to Nasim's capacity to privately interpret, and rework, the cultural idioms about Pashtun-ness, and masculinity in particular, that he had been enculturated into. Nasim is not only a passive element within his own cultural representational world, but acts as well as an agentic subject within it. He speaks from the standpoint of an individual who feels entitled to express and represent a different understanding of certain cultural idioms, and yet to maintain the cultural legitimacy to be called and considered a Pashtun— that is to say, to maintain within his own public image those diacritica that make him a Pashtun in the eyes of others, as well as his own (cf. Fredrik Barth [1969] on cultural diacritica). In other words, Nasim seemed to me not only to be shaped by the cultural representational world in which he is immersed, but also to maintain a degree of agency that makes him believe that his own self-image as the "good Pashtun," albeit different from the one many others endorse, has the legitimacy to be considered socially valid nonetheless, to be "good enough". In this way, Nasim holds the potential to shape his own cultural idioms, as much he is shaped by them. Yet, he vacillates between a sense of helplessness, and one of empowerment.

In 2012 and 2013, Nasim's life went through some considerable changes. He recovered from the loss of his NGO position and decided he would pass the bar exam as a defense lawyer. He succeeded, and in 2012 started representing inmates incarcerated in Nangarhar's provincial jails without legal guarantees.[1] Although the job did not pay much, and was fraught with problems stemming from difficult relations with the state authorities, he told me that he was enjoying his new activity and was doing better than in the past. He also started to move his wife and daughter from Jalalabad to Kabul

---

[1] Nasim explained that a high number among those who are incarcerated in Afghanistan come from the poorest and most disenfranchised sectors of Afghan society ("The rich and powerful, if they ever get arrested, don't stay in jail for too long, whatever their crime," he said to me). These men (and women as well) have hardly any idea of how the state legal system works. They are often not provided with a state defense lawyer, and are not made clearly aware of the reason and evidence for their incarceration. Many remain in prison for indefinite amounts of time, without ever going to trial. Instead of reclaiming their relatives' rights under Afghan law (which they ignore), the families of these inmates usually look for alternative methods to get them out of jail (like bribes, personal connections, etc.). The task of Nasim was to tour the jails of Nangarhar province and check on the status of those who were held there without any legal representation or whose cases had not been processed fairly. He would then offer to represent them for a low fee.

As an interesting sidenote, Nasim explained also that those who end up in the provincial jails have often already spent time in the jail managed by the NDS in Jalalabad (National Directorate of Security, the successor of KhAD, the intelligence agency of the Afghan Communist governments). The NDS screens for terrorist links and activities. Those who are released and transit into the provincial jails are implicitly considered to be low-profile criminal offenders. However, those who are kept in the NDS jail are completely off-limits to anyone, and are under the sole authority of the NDS interrogators and guards. It is not difficult to imagine the type of treatment they receive while in custody of the NDS.

during the summer months along with his grandfather. Shifting residence to cooler places during the harsh summer months of Nangarhar province is not an uncommon custom among residents of Jalalabad and the rural districts who can afford to do so. However, in the light of our conversations during the previous years, I could not help but consider his choice as motivated also by the need to satisfy some of his most compelling desires, as he expressed them to me over the years. At least during the summer months, he could join his wife and daughter in Kabul almost on a weekly basis, and live for a few days "more free," as he put it so often. I deemed it part of Nasim's tenacious inner working at constructing, and claiming for himself a self-image as a "good Pashtun" that could resonate as legitimate within the context of the cultural representational world of his environment, in spite of the unorthodoxy of his approach to it. Although still in part powerless to affirm his interpretation on how "to be good at being a Pashtun" at home in Jalalabad (to paraphrase Michael Herzfeld. Herzfeld 1985: 16), Nasim I believe started a process of self-acknowledgment that would necessarily, in due time, spill over in his home sociocultural milieu.

Shortly before I left the field in mid-2013, Nasim settled down with a new, permanent position as a legal adviser to an important Afghan government oversight agency. The shift, and the new responsibilities, made him visibly happy. He was glad, he told me, that he could actually make a difference to people's lives in Afghanistan. His ever-present pragmatism and inclination to a disillusioned realism made him add that he was still rather sure that after the pull out of the foreign troops the situation in the country would spiral for the worse. Yet, his new position, whose existence was guaranteed by the organization's close ties to the United Nations, made him sure that he could have a constructive role even in the case of a catastrophic, warlike scenario. Nasim's hiring by the Commission was facilitated by relentless activity on the part of his father, Niamatullah, who pulled all the strings he could in order to get his son the job. Nasim still had to go through a selection process, he told me, yet the many phone calls and meetings his father had on his be-half certainly served him well. It is also true that all the other candidates, most likely, had to pull their own strings as best as they could in order to maximize the chances to get hired. Yet these are the (accepted) rules of the game in Afghanistan, both in private and in state sectors. It is considered standard, even necessary, to proceed this way. *Andiwalay*, or the art of net-working, and getting recommended by someone powerful enough (literally, in Pashto, "friendship," in a euphemistic and almost romantic form, along the lines elaborated by Charles Lindholm [1982: 240–273]), is a crucial aspect of social life.

I was never surprised at the way Nasim's profile developed under my eyes, in the four years we were in personal contact with each other. The more I knew his father, the more I could appreciate the influence he had had on Nasim. Niamatullah had been a Khalqi, Nasim told me, that is to say a member of the Khalq faction within the PDPA (the Afghan Communist Party), opposed to their rivals, the Parcham faction. Khalq in the end lost to Parcham, which ruled the country with Moscow's blessing for most of the 1980s until the fall of the regime in 1992. Khalq, however, maintained its strongholds within the military and in rural areas (Arnold 1983: 37–51). It was a strongly Pashtun-based faction, and as such represented a Pashtun-nationalistic wing of leftist-minded intellectuals and activists. Their loosely Marxist-oriented beliefs were associated with a strong ethnocentric mindset. Such an unusual cocktail allowed the Khalqis to reject religion as a basis for social planning and public representation, but they maintained the conviction that Pashtun political and intellectual elites should continue to lead Afghanistan and its state apparatuses, as they had since the inception of the modern Afghan state in the late nineteenth century.

By the time I met him Niamatullah was not any longer the old-fashioned Khalqi he might have been in the 1980s (if at all). The lectures given by him in Pashto to his collaborators, whom he trained to go to the rural districts and implement the projects his civil society organization had won for itself, were all about the pros and advantages of democracy, free elections, decentralized administration, and citizens' rights, in line with the general principles that the Western international donors wanted to propagate among the Afghan population. Certainly, he seemed very passionate and genuinely convinced of the narrative he presented to his collaborators. I cannot say whether he was really so, or was simply executing the task given to him. To me, in private, he railed against tyranny, dictatorship, and the powerlessness of the people in the face of their political masters, declaring that the Soviet Union itself had turned into a dictatorship and that Communism was not a democratic system. At the same time, though, he remained convinced that the period of the presidency of Mohammed Najibullah, the last Communist president (albeit Parchami), remained a golden age in Afghan history. There was very little corruption, he said often, as state officials worked for the good of the community, keeping in mind the well-being of the population. Najibullah himself possessed nothing of his own when he died, other than a modest house in Kabul, and so did most of the other state officials, he recalled, who would have spent the rest of their life in jail if they had been caught stealing from the public coffers.

Whatever his current ideas on Marxism, socialism, and the sociopolitical structure these ideologies advocated, Niamatullah always seemed to me to have retained at least the Pashtun ethnocentric inspiration from his years in the state apparatus and the party. He never said anything, or commented in any ethnically inclined way, on figures of non-Pashtun ethnicity who appeared on the political stage during the years I spent in Afghanistan. However, and this did not escape my attention, he never proffered a single word in Persian when talking to people who spoke Persian as their first language. This might seem a trivial detail, yet in Afghanistan language politics have a great importance. Niamatullah knew Persian well, as all the educated Pashtuns generally do in Afghanistan. In fact, most ethnic Pashtuns who led the country in the past were, for historical reasons, Persian-speaking, and speaking good Persian was a requisite for working in the state administration then, as it is now. Niamatullah could understand everything a Persian-speaker would tell him. I found out this, to my surprise, during the few times he had to meet in his office with people who legitimately could not speak Pashto. He would listen to them speaking Persian, and then he would reply in Pashto addressing someone else in their group who could understand him and translate. I finally discovered he could also speak fluent Persian when he once had to speak on the phone to someone in Kabul who was not a Pashto speaker. Among his close collaborators there were young men who were born and raised in Jalalabad, but were of non-Pashtun ethnic descent, whose family language was Persian. He never spoke to them in Persian, and mostly they made the effort of speaking to him in Pashto. Sometimes, though, in the heat of a conversation, they would inadvertently revert to their first language (Persian) which Niamatullah understood, but stubbornly replied in Pashto.

It is very likely that the importance of Pashtun cultural values for Nasim, especially in the realm of masculinity and social respectability, have to do in good measure with Niamatullah's own formation, and the type of education he gave his only child Nasim. Nasim identifies with certain idioms of Pashtun masculinity (ghairat, land owning, the protection of one's "natural" rights) and disavows those that traditionally link pukhto with Islam. This may explain why Nasim manages to interpret his Pashtun-ness in a more malleable, "elastic," and inclusive fashion. A "secular" understanding of pukhto, as it were, which he might have inherited from his father's life trajectory, allows

Nasim to construe a self-image as a "good Pashtun" that diverges from the standard idiom that his cultural representational world has passed on to him. Nasim considers this self-image as a legitimate one both socially and culturally. Hence we saw, on the one hand, his resilience in choosing to stay and live in the place where he was born, taking care of his family's possessions, as any "good Pashtun" should do (recall what Rahmat Waraki thought about this), and on the other, his confidence in thinking that he can remain a "good Pashtun" even when he treats his wife in an unconventional way, or when he hopes for a change in the much too harsh and violent implementation of pukhto in the everyday life.

Nasim is the representative of a new generation of urban, educated Pashtuns, who are however still deeply immersed in the ethnocultural realities of their milieu, and do not reject them. Paradoxically, he might appear like an updated version of the (now rare) ethnically proud Pashtun who harbors "progressive" sentiments and expresses them through the language of an idealistic political philosophy, such as the Marxism-Leninism of the 1970s and 1980s. Whereas his father found his language through his partial Soviet education and the contemporary political Afghan narrative, Nasim today finds his own in the massive inputs that come to Afghanistan from Western political, developmental, and intellectual sources. The socialist propaganda that Niamatullah endorsed in order to give shape to his worldview in the 1980s has been replaced for Nasim by a contemporary one based on "Western" principles of democracy, gender equality, civil liberties, and political representation. Nasim's persistent self-image of Pashtun-ness, of representing the "good Pashtun," can be adjusted to these contemporary ideological inputs and be rendered "modern," just as his father's Pashtun-ness could proclaim its "modernity" through aligning itself with a socialist movement that claimed to embody the modernity of the Afghan people.

Wahid

Wahid is one of the most remarkable characters I interacted with during my fieldwork research. Unfortunately, I never had the possibility of working with him in the deep and thorough way I did with others (although I would have much liked to do so) because he was a highly focused, tenacious, and goal-oriented young man. He realized fairly soon that I could not provide the help that he was looking for, which resulted in him losing any interest in spending too much time with me. After the first months, in 2010, I could meet him only occasionally, although we stayed in touch until the end of my fieldwork. His goal was to leave Afghanistan as soon as possible and to gain a foothold in the United States. Wahid's profile was remarkable in that he was the only one Pashtun man I met who uncompromisingly, vocally, and unapologetically rejected his Pashtun-ness, his religion, and religion in general. His "ideological" positions were so extreme from an Afghan standpoint that he could have been called a "revolutionary" if he had cared for political activism. A student in the English department of Nangarhar University, in Jalalabad, Wahid spoke spectacular English with a strong American accent. He was born and raised in Jalalabad, but both his parents came from Khogiany, a mountainous district in Nangarhar province notorious for its traditional and conservative culture (and its combative inhabitants), which was often used in everyday conversations by other Pashtuns in the province as the stereotype of backward and "wild" people. A probably apocryphal belief was that Daoud Khan, the last president of Afghanistan before the Communist takeover, had readied a plan to literally fence off the entire district, so to keep the "savages" away from the rest of the "civilized" people. In 2010, when I first met Wahid, he was in his last year of college. In August 2012, when I started my last research stint in Afghanistan, he had graduated, and was going through the selection for a Fulbright scholarship in the United States. I was still in the field when, in mid-2013, he finally fulfilled his dream, and left to the United States in order to get a master's degree in development studies within the Fulbright program. The only problem with such an idyllic picture is that all Fulbright scholars commit to return to their country after graduating in the United States, in order to apply their enhanced knowledge to their country's benefit. It is part of the deal, and there is no way out of it. Wahid will be obliged to go back to Afghanistan after completing his degree, unless he goes to another country.

The way I first met Wahid is indicative of the character he represented. A loose acquaintance of mine in Jalalabad, in summer 2010, told me that a classmate of his from university spoke very good English, and he thought he might be interested in talking to me. I happily accepted the offer and we went to meet his friend Wahid at the university playing field. Wahid was intent in a basketball game with other young Afghans. At the end of the game we were introduced. Besides surprising me with the quality of his spoken English, he explained to me that basketball was a sport that had been introduced vigorously by the few foreigners who, over the years, had lived in Jalalabad, and taught English, either privately or at the university. These foreigners were usually American or Canadian. I would realize later (after meeting two of them) that they were invariably Christian Evangelical missionaries. Proselytizing is an illegal practice in Afghanistan, as it had been for a long time, and the missionaries usually covered up their main activities by teaching English (for free) or getting employed in a foreign NGO working in the area. In his goal-oriented "fury," Wahid had befriended a few of those living in Jalalabad and used to spend a sizable amount of his time together with them. After the last missionary left, in mid-2012, he kept frequenting a group of US contractors who worked for the Department of Defense on satellite imaging and Internet connections in the region. They resided in a heavily fortified guesthouse for foreigners only on the outskirts of Jalalabad. In 2012, Wahid took to spend a few evenings every week at this guesthouse, which my wife and I visited once. I doubt the place escaped the attention of the local people, who very likely understood that foreigners were living and working inside the compound. Wahid's acquaintances certainly helped his knowledge of the language and kept him emotionally close to his dream of reaching the New World, by helping him network in order to facilitate a future move to the United States. However, he underestimated the social consequences of his too evident frequentations and in the end he paid a high price for it.

Wahid was fatally attracted to anything that was American. He had chosen the United States as his own lifeboat out of Afghanistan. By age twenty-two in 2010, he had spent his adolescence in a province militarily occupied only by US troops. Though we never discussed the presence of the US soldiers all around him, it is easy to imagine how such a state of facts might have acquired a positive, even alluring, feature for Wahid's life plans. Not that the materialization of the (armed) "Other" in town in the wake of the demise of the Taliban should be considered the only cause for a whole structure of meaning, dreams, and wishes, that Wahid constructed during his adolescence. It might well be that this "Other" came just at the right time to embody and give concreteness to ideas and emotions that Wahid might have been (whether consciously or unconsciously) developing autonomously. This is roughly the version by which Wahid himself interpreted his adolescent trajectory. In January 2011, he told me:

> I started questioning things since the very beginning . . . since I can remember,
> in school. I would ask myself why this and why that, especially about religion.
> I would sometimes ask the teacher in school about things like fasting during

Ramadan or heaven, and Mohammed . . . he could not give me any explana-
tion, he just kept on repeating what we found written in our books, without
actually explaining anything . . . the whole school was like that . . . all the
subjects, just repetition, no thinking . . . useless, completely useless . . . a bunch
of idiots, that's what they were. I started reading things by myself, I wasn't even
listening anymore to the teacher in class. . . . School in Afghanistan is good for
nothing. . . .

With our relationship continuing over time, his defenses also came down little by little, and he could afford to talk more openly about "incriminating" topics. Right after Ramadan 2011, while walking back to my apartment from the house of a common friend in Jalalabad, we finally discussed religion:

WAHID—*I don't give a damn about religion . . . it's all nonsense to me. . . .*

ANDREA—*You mean Islam? Yeah, I had the impression you might not be the*
perfect kind of pious Muslim. . . .

WAHID—*No, not Islam, I mean religion in general. . . . I really don't care about*
any of the official religions . . . it's all stories for the uneducated people. . . .

ANDREA—*Ha, that's interesting . . . you know, I was starting to think that you*
might have secretly converted to Christianity, with all the time you spent
with the missionaries. . . .

WAHID—*No, no, are you kidding? No, I don't care about that either. . . . I mean,*
not that they did not try, you know how they work, they do try sooner or
later, they do talk about Jesus . . . but I never picked up on that, I just did not
care . . . they kept on spending time with me because probably they never lose
completely hope in someone to convert. . . . I just wanted to keep practicing
English, you know that. . . .

ANDREA—*Well, OK, but all the more if you did not convert, you have to be*
careful not to spend too much time openly with them, you are putting your-
self in a risky position. . . . I mean, also the guys at the guesthouse, people
know what they do . . . it can become dangerous for you to hang out too often
with them. . . .

WAHID—*I know, don't worry, I know that I have to be careful . . . so far eve-*
rything is under control . . . don't worry. . . . I only want to get out of this
place. . . .

He also explained to me that he does not fast during Ramadan and does not go to mosque. He does pray when in public though, when others do the same.

This he perceived as hypocrisy on his part and that of many others. He justified this by the extreme intolerance of the environment. With an amused tone, he described to me all the logistical gymnastics that he goes through in order to eat secretly, away from anybody's gaze, during the holy month of Ramadan. His parents do not know he does not fast, he told me, and he has to hide food here and there around the house in order to eat whenever nobody is looking. The whole ordeal makes him feel "trapped, like in jail . . . and very sad," he added.

Unfortunately, Wahid, at a certain point, must have overstepped, or something must have gone the wrong way for him. When I came back to Afghanistan for the last, long-term section of my fieldwork, in summer 2012, I called Wahid on the phone upon arriving in Kabul, thinking we would arrange to see each other as soon as I would get to Jalalabad again. Yet he was already in Kabul, where he had relocated a few weeks earlier. In early 2012, he and a close friend of his, who shared his same acquaintances, suffered personally the consequences of the increasingly deteriorating security situation in the province. Somehow a rumor spread that Wahid and his friend had converted to Christianity. Suddenly posters with their two pictures, a description of both, and an inflammatory text appeared on the walls of the university in Jalalabad. The text described how the two had for years been close to Christian preachers and missionaries in town, and how they had been working for US organizations. It stated that Wahid and his friend had become Christians and must pay with their lives, as shari'a law requires in cases of apostasy (i.e., abandoning Islam for another faith). In addition, Wahid told me that the house of the only one missionary left in Jalalabad, where he lived with his wife and two young daughters, had been bombed at the beginning of the summer, although fortunately nobody was injured in the attack. My wife and I had been guests of the family for dinner, in the same house, in summer 2011. They reportedly left Jalalabad after the attack.

After updating me on the situation by phone, we eventually met in Kabul a few hours later. Wahid had left precipitously Jalalabad for Kabul after the posters appeared on the university walls, and he was living alone in an undisclosed location. When we met, in a Kabul restaurant, he was clearly in distress. He was angry at what had happened to him. He was angry at his people.

I hate these people. . . . I hate Pashtuns. . . . They're only a bunch of ignorant savages . . . with their customs, and their rules . . . what kind of life is this? They're like fanatics, about religion, about their way of life . . . but they are just ignorant. . . . I have no place among these people. . . .

ANDREA—*Well, you are a Pashtun as well.* . . .

WAHID—*Yes, I know, but I have nothing to do with them, I am not like
them.* . . . *Yes, I am a Pashtun because I was born here and I speak Pashto,
but I am a different person, I don't live like them.* . . .

He told me that his family had also paid a toll because of his lifestyle. Both
his mother and father had been dismissed from their jobs (the mother as a
schoolteacher, the father in a construction company), and had been harassed
and marginalized by their neighbors. His mother did not leave the house any
longer, while his father was still in a state of confusion, he said, not quite sure
about what they should do. Yet, Wahid did not seem too worried about the
predicament in which his public behavior had plunged his family. He was well
under way through the application for a Fulbright scholarship to the United
States, and it seemed that the high hopes he held in this regard were enough
to keep his morale up. I would have frankly expected some more concern for
the plight of his parents. Being the mother and father of a suspected apostate
may bring fatal consequences in a Pashtun milieu, all the more in the light
of the recent spike in religious radicalism. The life itself of his parents might
have been in danger as long as they remained in Jalalabad. I suggested that he
urge his parents and brother to move to Kabul with him, at least temporarily,
and help the storm die down for the time being. He explained to me that his
father was resistant to relocation, and that they would try to continue living
in Jalalabad as long as possible.

All in all, however, it seemed to me that Wahid was somewhat oblivious
to the responsibilities that he had in the dramatic development of the situation. I had warned him over the years several times of the possible backlash that his all too patent and public association with figures who people
knew had a "suspicious" background might have in the community. Wahid
always dismissed my warnings by professing his awareness of the situation,
and underlining his efforts at defusing the dangers. Yet I could not help but
think that his behavior was rather reckless, within the volatile and precarious
conditions of security in Nangarhar province during those years.

It was obvious to me that Wahid had a very strong desire (and ambition)
to leave Afghanistan. This was a vital priority in his life plan and he was
ready to take any risk in order to ensure a higher chance of succeeding in
the task, even if this would have meant jeopardizing the safety of his family
in the process. Many other Afghan friends, as Wahid, had expressed to me
the burning desire to leave their country, a place where they saw no viable

future for themselves. Some had even taken the extreme measure of illegally emigrating to Europe, only to face the grim fate of being sent back after being caught by the authorities of one of the European border countries. However, choosing to illegally emigrate was an accepted, "culturally legitimate" way of finding a way out for oneself, which entailed financial sacrifice in order to find the necessary money to organize the trip, personal risk while undertaking the enterprise, and, above all, no social consequences for any of the people left behind—all features that were consistent with a historically established tradition of "manly" Pashtun responses to familial hardships and social crises (see Nichols 2008).

The way Wahid chose to find a way out of Afghanistan, however, was anything but "culturally legitimate." He was trying to take advantage of connections and networking that he created through the years with US and Canadian nationals, who were in Afghanistan to conduct activities that were not only prohibited by law, but notoriously considered offensive and reprehensible by the vast majority of Afghan Muslims. Sure enough, he was applying the concept of "andiwalay" (exploiting connections) in a very culturally coherent and effective way, but he was doing it with the "wrong" people. His excessive public exposure alongside foreign individuals of a "suspicious" kind will have certainly facilitated his plan to leave Afghanistan, yet the price that his family paid for it was dire. Wahid in the end left for the United States on a Fulbright scholarship, which made his strategy fully successful, but his family was left behind to deal with the social consequences of having a son who was thought by most to have really converted to Christianity. Furthermore, the fact that he would spend the following two years in the United States would reinforce and corroborate the suspicions of many about Wahid having abandoned Islam.

What always fascinated and intrigued me about Wahid was the puzzle of how a fairly traditional and rural Pashtun environment, such as the one he had lived in, split between the city and Khogiany district, could have produced such a sanguine, vitriolic, and rebellious character, so vociferously opposed to the cultural representational world, and social arrangements that his own milieu upheld and performed. The lack of interest, on the part of Wahid, to explore together with me the potential reasons for his personal development prevents me from even attempting to elaborate a possible explanation. Certainly, a powerful independent variable in Wahid's life's equation must be represented by the massive influx of Western ideas and allegedly universal social principles that poured into the country after the demise of the Taliban, in the guise of NGO workers, United Nations programs, political advisers, and military personnel. Wahid spent his adolescence in the midst of this ideological turmoil, and it would be unrealistic to imagine that it did not influence profoundly the imaginary, and self-representation, of a developing young man. To what degree, and in which specific directions, such influence interacted with Wahid's private world, and social dynamics, is not possible for me to discern.

To be sure, Wahid was probably "more Pashtun" than he would have liked to admit, at least in certain regards. For example, he was the one who made great fun of me for not having sexual relationships other than with my wife, even before the wedding, after agreeing to be her partner. Not only could he not fathom why I wanted to "eat every day the same rice," as he put it, but he did consider within the legitimate prerogatives of a man (of whatever background) that of having extramarital relationships with whomever he wanted. My wife, obviously, was not accorded by him the same prerogatives. Wahid was well aware that adultery is frowned upon, and punished, in Islam, for women as well as for men. "That's how it goes here, everybody does it," he commented, in justification of his position. Thus, in spite of the "hyper-Westernism" that he displays in admiring life and customs of "the West" (which for him means mostly the United States), and condemning those of his own sociocultural milieu, Wahid in certain respects is still very much a man of his time and place—especially when he has something to gain from it, as it seems. And it could not be otherwise, after all.

Nevertheless, whatever the contradictions and inconsistencies, the very fact that such a (by and large) iconoclastic and nonconformist figure came to the public surface (so much to become exposed publicly on the walls of his university campus) speaks to the great dynamism and creativity that the Pashtun sociocultural environment, despite its reputation for stability and traditionalism, acquired in recent years (at least), in the wake of the international intervention in the Afghanistan.

Conclusion

The search for "meaning" within a shared psychic reality

In the early 1930s, Edward Sapir had this to say to his students at Yale, during a series of lectures that Judith Irvine has reconstructed for us from the students' class notes: "The relation between personality and culture—that is, on the level of observable behavior, between behavior expressing the personal concern of the individual and behavior expressing cultural forms—has become my obsession" (Sapir 2002: 176). To correctly interpret this passage, one must realize that what Sapir called "personality" was much closer to what today we call "subjectivity" than to the meaning most anthropologists of his generation (and even later ones) gave to the concept of "personality." For Sapir, "personality" is not the end product of a process of maturation that peaks sometime between adolescence and early adulthood, and stays invariant for the rest of the individual's life (and that can thus be neutrally observed, dissected, and "analyzed"). Rather, it is an accretive and ever dynamic process that takes incremental advantage of the import and meaning that interpersonal relations, in social life, bring to the individual's psychological growth (he thus incorporated the intuitions of Harry Stack Sullivan's theory of dynamic psychiatry). Likewise, for Sapir, culture was not the *cloture heréditaire*—the hereditary constraints (see Fassin 2000)—that many of his contemporary colleagues believed it was. The individual is for him not a *homo culturalis*, a cultural animal "all the way down" (to paraphrase Clifford Geertz).[1] On the contrary, culture for Sapir was something that each individual would rework and subjectivize throughout his or her life. Culture, additionally, was also inclusive of the political and social developments of the

---

[1] It is worthwhile here to quote Adam Kuper to the same effect: "Unless we separate the various processes that are lumped together under the heading of culture, and then look beyond the field of culture to other processes, we will not get far in understanding any of it.... We all have multiple identities, and even if I accept that I have a primary cultural identity, I may not want to conform to it.... If I am to regard myself only as a cultural being, I allow myself little room for maneuver, or to question the world in which I find myself" (Kuper 1999: 288).

Crafting Masculine Selves. Andrea Chiovenda, Oxford University Press (2020). © Oxford University Press.
DOI: 10.1093/oso/9780190073558.001.0001

society in which one lived (and in this sense, coming also close to what Clyde Kluckhohn thought of culture; see Kluckhohn and Murray 1948).

Against this background, I can say without fear of being misunderstood that this book tried to pick up, through a clinical ethnographic methodology and the support from leads in contemporary psychoanalytic and anthropological theory, an intellectual and ethnographic thread that was left pending by Edward Sapir at the time of his death in 1939.

The previous ethnographic chapters have shown, I hope, that each person's subjectivity does not exist and function as a "stand-alone" and impermeable microcosm, with its endogenous and autonomous mechanisms. Accordingly, a subjective system of psychic functioning cannot work in isolation: it has to "attach" to some aspect of outside reality in order to operate. Psychic processes have to be "triggered," so to speak, by external contingencies. Such external contingencies, cultural and social in kind, trigger idiosyncratic processes in different individuals, and in so doing produce unique subjectivities.[2] The cultural lens allowed by a specific idiom, masculinity— which is of paramount importance for the social life of Pashtun communities in Afghanistan—was the "trigger" that I chose to study my informants' psychic dynamics. We have seen, through the discussion of the psychodynamics of the main characters of these chapters, that common, shared patterns of psychic functioning are in fact enacted and pragmatically performed by each individual and are totally context-specific, responding to precise environmental cues. The inner conflicts that Pashtun cultural idioms of masculinity engender in Pashtun (male) individuals, the social arrangements and peculiar forms of interdependent relationships that they produce between those same individuals, and the emotional ramifications they obtain in them all impact profoundly the way shared patterns of psychic functioning are played out in real life by my informants, in their own idiosyncratic ways. To this, we have also to add the specific interpersonal relations that political developments in Afghanistan at the time of my fieldwork created for my informants (mainly, a long-lasting state of severe conflict, the occupation of the country by yet another external power structure—the US-led coalition since 2002—and the

[2] In this regard, Steven Parish has written that "[T]he mind here does not—mindlessly—duplicate external cultural forms. What was external, part of culture and society, becomes internal, part of mind and self, but is transformed. This transformation creates *distance between culture and self*, but links them in a dynamic, dialectical relationship. This cultural self is not 'programmed' by culture, but is *primed to find meaning in culture*" (Parish 1994: 294, emphasis added).

creation of an Afghan state apparatus that many Pashtuns considered all but illegitimate and unrepresentative).

As Jerome Bruner pointed out, the crux of the issue lies in "meaning" (Bruner 1990). It is important here not to equate the function of a mechanism with its supposed explanatory power. An anthropologist and a psychiatrist, discussing projective identification and its (mis)use by clinicians to explain a vast and diverse gamut of psychological phenomena, famously wrote that "projection is a mechanism, not an explanation" (Littlewood and Lipsedge 1989: 29). It is the "meaning" that individuals give to the psychic processes they (consciously or unconsciously) make use of, that leads us closer to a fuller explanation and understanding of the reasons why they do so, and, more broadly speaking, why they behave (and "believe") in a certain manner. My interaction with the characters of this book showed that "meaning" can hardly be extracted only from the study of intrapsychic dynamics and endogenous mechanisms, or from the analysis of shared cultural idioms.[3] "Meaning," as Clifford Geertz understood in his quasi-Weberian approach to anthropological theory, is created by each individual through privately interpreting and reworking the cultural material and social constraints in which he or she is immersed (in fact, Geertz was much more of a "psychologist" than he might have liked to admit). However, such acts of interpretation and "manipulation" ("acts of meaning," in Bruner's words) are attained only through the workings of psychic processes and dynamics, whose general patterns are common to humans in general, but whose manners of utilization are peculiar to each individual, and are context-specific. It is in this sense that "meaning" becomes really "personal," for it couples cultural and social material with idiosyncratic psychic processes (the "subjectification of cultural symbols," as Gananath Obeyesekere termed this phenomenon. Obeyesekere 1981. See also Hollan 2000). And it is also following this path that each individual obtains in turn the possibility of affecting his or her sociocultural milieu as an actor of change.

In this regard, British psychoanalyst Farhad Dalal writes: "Each of us, as a particular individual, is born into a pre-existing society constituted by a multiplicity of overlapping and conflicting cultures. The cultures

---

[3] Sure enough, what developmental psychologist Jerome Kagan calls "temperament" (an idiosyncratic and unique aspect of one's own personality configuration) has also a crucial role in how individuals respond to certain specific environmental cues. However, our understanding of what "temperament" is, its origins, and its impact on one's psychological profile is still too vague to be taken into consideration as an acting variable in our investigation of human behavior (Kagan 2013).

themselves, as well as the relationships between cultures, are constituted by power relationships. As we grow and develop, we imbibe, of necessity, the pre-existing cultural forms, habits, beliefs, and ways of thinking around us. These introjections are not taken into an already-formed self; rather, they come to *contribute* to the formation of that self" (Dalal 2006: 145, emphasis in original). Dalal's position does not give way to any sort of social determinism: "To think in the way described does not mean denying the existence of individuals, each with his or her own unique sense of self, nor denying that all are biological beings in bodies" (Dalal 2006). In fact, Dalal's position is only complementary to Joao Biehl's, Byron Good's, and Arthur Kleinman's, when they aptly argue that "in a 'world in pieces,' older notions of the subject who is cultural 'all the way' seem inadequate" (Biehl, Good, and Kleinman 2007: 8).

Furthermore, my emphasis on the interpersonal and relational aspects of my informants' lives does not imply that conflict, competition, and aggression should be seen only as epiphenomena, vis-à-vis a supposedly more deeply rooted inclination toward social cooperation. My point is that the very concepts of relatedness and interdependence include necessarily within themselves competition, friction, and possibly conflict.[4] Conflict, whether social or interpersonal, is but one aspect of the relatedness on which human lives are premised. It is a relational, interpersonal, ecological rationale for the emergence of conflict (as opposed to the endogenous, intrapsychic rationale peculiar to the vast Freudian and ego-psychological tradition in psychodynamic studies). A considerable part of the material I gathered from my informants is pervaded by such dynamics, which they handled and managed both consciously and unconsciously, and with which I tried to interact accordingly.

The very same material leads us to a complementary way to look at conflict, and its roots—that is, through a conceptualization of "power." All relationships of relatedness and interdependence are inherently informed by "power," which structures relatedness and interdependence along lines of domination and subordination—whether material or symbolic. Norbert Elias maintained a useful and nuanced conception of "power," eschewing essentialist and reductionist views of it. Power, he writes, "is not an amulet

---

[4] Many contemporary developmental psychologists and psychoanalysts have come to interpret the roots of human relatedness in these terms as well (among the most notable ones, see Stern D. 2001 [1985], Fonagy 2001, Mitchell 1988).

possessed by one person and not by another; it is a structural characteristic of human relationships—of *all* human relationships" (Elias 1978: 74, emphasis in original). Power entails, among other things, conflict and antagonism. For this reason, the concept of relatedness fully comprises instances of social and interpersonal conflict as well. Again, dynamics of domination and subordination are omnipresent in the material that my informants presented to me, whether with regard to masculinity, violence, or personal agency.

When we acknowledge that the individuals whom we study are the locus (the agents) for the creation and negotiation of cultural and social material, we have also to realize that analyzing the (publicly) unseen, unspoken, even unconfessed inner dynamics that these individuals harbor takes us one step closer to understanding the "mechanics" through which such cultural and social material may change over time, diachronically. I hope I showed convincingly enough that the uncertainties, fears, conflicts, contradictions, achievements, fantasies, rejections, rebellions, and endorsements that my informants displayed in their relationship with me, and in their daily life behaviors, sow the seeds of change in the communities where they live. Umar and his shift from radical militant to "enlightened" pious Muslim; Kamran and his stubborn insistence in a more "Islamic" way of being a good Pashtun; Rohullah and his effort (not always successful) to leave behind the violent side of his being a Pashtun nar (manly man). These are but some of the instances that I have analyzed in detail, in which single individuals operate powerfully on their social and cultural milieu, and by way of their public example constitute a new paradigm, or a new model, that may inspire others as well. The pace of cultural and social change is unfortunately too slow for an immediate observer to obtain any inescapable proof of such dynamics. Yet it seems likely to me that when the number of individuals embracing a certain attitude slowly and incrementally reach a critical mass, then models will be replaced and new paradigms will take hold. Fredrik Barth (1966) and Anthony Wallace (1956)—in my opinion two of the most lucid and perceptive authors who dealt with this issue—envisioned change in these exact terms. We need to understand the individual to reach a fuller perspective on how, and why, change may eventually happen.

From the standpoint of the individual psychological dynamics with which this book has been primarily concerned, we have seen the emergence of a multiplicity of subjectivities, of self-images, that my informants gave birth to and managed, in order to cope with changing historical circumstances that were greatly affecting their personal lives. Contemporary psychological and psychoanalytic literature, which I have referenced extensively in the previous chapters, have now acknowledged that the reification of a concept such as "the self," as if in the guise of a real entity clashing with other inner entities (the ego, à la Freud, and ego-psychology authors), or a monolithic structure that could undergo processes of fragmentation and even disintegration (à la Heinz Kohut, and subsequent self-psychology), is not a satisfactory way to deal with what people (and analysts) would experientially term "self." I feel that with "self" we must now mean a set of different subjective states, which coexist, complement each other, and compete within an overarching self-representational world maintained by the individual. Each set of subjective states gains its meanings, and symbolic material, from the private management of the social and cultural context in which the individual exists, and yet is "operated" through common psychic dynamics and mechanisms that all humans share.[5] The idea of the existence of a coherent self, or ego, must be replaced by the notion of the *illusion* of the existence of a coherent self (as Katherine Ewing suggested long ago), which in reality represents the coalescence of multiple, shifting states of subjectivity. The fact that it should be considered an illusion does not take anything away from the fact that it is nonetheless a *necessary* illusion, in order for the individual to maintain a productive, functional, and, in the end, healthy psychological equilibrium. I think that the material that my informants presented to me, and the way it developed through their intersubjective encounter with me, might corroborate such a theoretical position.

Contemporary psychological anthropologists, like Rebecca Lester (2017), Jeanette Mageo (2015) and Tanya Luhrmann (2012) have produced highly sophisticated and nuanced investigations of individual subjectivity, focusing phenomenologically on culturally structured self-representations and idioms of selfhood. I privileged an intersubjective exploration of modes of psychic functioning and unconscious dynamics, immersed as they are in a complex "cultural broth" (to use a phrase coined by Sudhir Kakar). In fact, alongside Byron Good, I believe that "the language of subjectivity signals

---

[5] Malcolm Owen Slavin and Daniel Kriegman write that "[W]hat we inherit (as unique individuals, and in our shared, universal inheritance as a species) are largely internal structural and dynamic features (ways of organizing and disguising experience, sensitivities, vulnerabilities, etc.) that through interaction with a given social and cultural environment will be manifested overtly (behaviorally) in a whole range of different ways" (Slavin and Kriegman 1992: 76 fn.10). To this they add that "we would expect that a major feature of the universal deep structure [of the human psyche] must consist of *mechanisms or strategies* for dealing with inevitable individual variation" (77, emphasis added).

a complex, psychological understanding of the subject, one for which current psychoanalytic conceptualizations and methods are relevant" (Good 2012b: 517).

My informants were socially and culturally functional individuals (i.e., not subject to any major and crippling psychological disturbance or psychiatric disorder), men who evidently managed to work their subjective states toward the overarching sense of coherence they needed. We have seen how their efforts at coherence were not devoid of contradictions, conflicts, and inner suffering. Nevertheless, as individuals who displayed a reasonably healthy psychological state, they were in the position of bringing their shifting and competing subjective states into some sort of internal cohesiveness.

An important corollary of this approach to subjectivity is represented by the aspect of individual agency, which also looms large throughout the book. I hope that the narration and analysis of my informants' diverse vicissitudes and predicaments showed that, far from being individuals at the mercy of their own cultural schemata and social arrangements, they managed to maintain (to various degrees) a productive and healthy contact with the subjective state that at any given moment represented their "perceptual truth" (what Wilfred Bion calls being "relatively truthful" to oneself. Bion 1982: 8). In the paradigm that I have proposed in this book, this happens when the shifting subjective states, of which the "illusion" of a coherent self is composed, are kept in productive communication with each other. In turn, this is achieved by a positive use of (unconscious) dissociative processes, whose culmination in conscious conflicts (i.e., the emergence into awareness of the dissimilarities between different subjective states) provides the material for psychological growth and emotional advancement (see also Bromberg 1996, 2003). I believe that the evidence presented by my informants may be interpreted, and understood, through acknowledging such phenomena.

These processes, when successfully achieved, allowed my informants to "sense" where their "individuality" lies vis-à-vis the constraints imposed on them by their sociocultural context. We have also seen that the interconnectivity that links to each other the lived experiences and emotional perceptions that my informants share with the people who surround them more closely, are part and parcel of their "authentic" selves (see Slavin and Kriegman 1992: 83–106), as a functionally adjusted element of their behavioral environment (Hallowell 1955). In a sociocultural context such as the Pashtuns', the self of each individual (i.e., his or her states of subjectivities) is also partially made up of aspects of the selves of other significant ones, as

Suad Joseph also intuited (Joseph 1999). It would be fallacious, I argued, to posit *tout court* any pathological side to this state of facts.

From a broader sociocultural and political perspective, I have contended that the very cultural idiom for masculinity (or rather, *masculinities*) among Afghan Pashtuns has undergone profound modifications over the past thirty-five years, due to continuous conflict within the country. Those informants who were born before the series of conflicts began were unanimous in reporting the perception that the boundaries that defined who was a nar (a manly man) had dramatically changed. They recognized that certain excessively violent and abusive behaviors, which were deemed a necessary evil during the war, had slowly become routinized and institutionalized in the absence of peace. The moral "state of exception" first applied during the revolt against the Communist governments continued during the civil war and the Taliban regime. The exception became the rule. The expression of one's own masculinity and manly worth premised on sheer force, ruthlessness, and violence became a new paradigm for a legitimate and culturally appropriate masculinity. I interpreted such a phenomenon through the lens of Fredrik Barth's transactional theory, which I believe fits the dynamics of change that Pashtun society in the southeast of Afghanistan was subject to. Just like Anthony Wallace's understanding of the change brought about by religious revitalization movements during times of profound social and cultural disruption (Wallace 1956), Barth's analysis stresses the role of the individual as the main actor of change, whether advertently or inadvertently.

Bibliography

Aase, Tor, 2002. *Tournaments of Power: Honor and Revenge in Contemporary World*, Ashgate Publishing Ltd., Aldershot, UK.

Abu-Lughod, Lila, 2013. *Do Muslim Women Need Saving?*, Harvard University Press, Cambridge, MA.

Agamben, Giorgio, 2005. *State of Exception*, University of Chicago Press, Chicago.

Ahmed, Akbar, 1976. *Millennium and Charisma among Pathans: A Critical Essay in Social Anthropology*, Routledge, London.

Ahmed, Akbar, 1980. *Pukhtun Economy and Society: Traditional Structure and Economic Development in a Tribal Society*, Routledge, London.

Ahmed, Akbar, 1991. *Resistance and Control in Pakistan*, Routledge, London.

Anderson, Jon, 1975. "Tribe and Community Among Ghilzai Pashtun," *Anthropos*, 70, pp.575–601.

Anderson, Jon, 1978a. "There Are No Khans Anymore: Economic Development and Social Change in Tribal Afghanistan," *Middle East Journal*, 32, 2, pp.167–183.

Anderson, Jon, 1978b. "Introduction and Overview," in Anderson, Jon, and Richard Strand, eds., 1978, *Ethnic Processes and Intergroup Relations in Contemporary Afghanistan*, Asia Society, New York, Occasional Papers, n.15, pp. 1–8.

Anderson, Jon, 1983. "Khan and Khel, Dialectics of Pakhtun Tribalism," in Tapper, Richard, ed., 1983, *The Conflict of Tribe and State in Iran and Afghanistan*, Croom Helm, London, pp. 119–149..

Anderson, Jon, 1984. "How Afghans Define Themselves in Relation to Islam," in Shahrani, Nazif, and Robert Canfield, eds., 1984, *Revolutions and Rebellions in Afghanistan: Anthropological Perspectives*, University of California, Berkeley, pp 266–287.

Arnold, Anthony, 1983. *Afghanistan's Two-Party Communism: Parcham and Khalq*, Hoover Institution Press, Stanford University, Stanford.

Aron, Lewis, 1991. "The Patient's Experience of the Analyst's Subjectivity," *Psychoanalytic Dialogues*, 1, pp.29–51.

Asad, Talal, 1972. "Market Model, Class Structure, and Consent: A Reconsideration of Swat Pathan Political Organization," *Man*, 7, pp.74–94.

Asad, Talal, ed., 1973. *Anthropology and the Colonial Encounter*, Humanities Press, New York.

Barfield, Thomas, 2008. "Culture and Custom in Nation-Building," *Maine Law Review*, 60, 2, pp.347–375.

Barfield, Thomas, 2010. *Afghanistan: A Cultural and Political History*, Princeton University Press, Princeton, NJ.

Barfield, Thomas, 2014. "Weapons of the Not So Weak in Afghanistan: Pashtun Agrarian Structure and Tribal Organization," in Johnson, Thomas, and Barry and Scott Zellen, eds., *Culture, Conflict, and Counterinsurgency*, Stanford University Press, Stanford, CA, pp. 95–119.

Crafting Masculine Selves. Andrea Chiovenda, Oxford University Press (2020). © Oxford University Press.
DOI: 10.1093/oso/9780190073558.001.0001

Barth, Fredrik, 1959. *Political Leadership among Swat Pathans*, Athlone Press, University of London.

Barth, Fredrik, 1966. *Models of Social Organization*, The Royal Anthropological Institute, London.

Barth, Fredrik, 1969. *Ethnic Groups and Boundaries: The Social Organization of Culture Difference*, Little, Brown, London.

Baryalay, Haroon, 2005. *The Jirga as a Dispute Resolution Body and the Case for Its Official Recognition by the State*, LLM Thesis, Harvard Law School, Cambridge, MA.

Baryalay, Haroon, 2006. *Jirga Theory and Reform Proposal*, LLM Research Paper, Harvard Law School, Cambridge, MA.

BGPSG, 1998. "Non-Interpretive Mechanisms in Psychoanalytic Theory: The 'Something More' Than Interpretation," *International Journal of Psychoanalysis*, 79, pp.903–921.

BGPSG, 2007. "The Foundational Level of Psychodynamic Meaning: Implicit Process in Relation to Conflict, Defense, and the Dynamic Unconscious," *International Journal of Psychoanalysis*, 88, pp.1–16.

Biehl, Joao, Byron Good, and Arthur Kleinman, 2007. "Introduction," in Biehl, Joao, Byron Good, and Arthur Kleinman, eds., *Subjectivity: Ethnographic Investigations*, University of California Press, Berkeley, CA.

Billé, Franck, 2015. *Sinophobia: Anxiety, Violence, and the Making of Mongolian Identity*, University of Hawai'i Press, Honolulu.

Bion, Wilfred, 1962. *Learning from Experience*, Basic Books, New York.

Bion, Wilfred, 1967a. *Second Thoughts: Selected Papers on Psycho-Analysis*, J. Aronson, New York.

Bion, Wilfred, 1967b. "Notes on Memory and Desire," *Psychoanalytic Forum*, 2, pp.272–273, 279–280.

Bion, Wilfred, 1982. *The Long Week-End, 1897–1919*, Fleetwood Press, Abingdon, UK.

Blok, Anton, 1981. "Rams and Billy-Goats: A Key to the Mediterranean Code of Honor," *Man*, 16, 3, pp.427–440.

Boesen, Inger, 1980. "Women, Honor, and Love: Some Aspects of the Pashtun Women's Life in Eastern Afghanistan," *Folk*, 21–22, pp.229–240.

Boesen, Inger, and A. Christensen, 1982. "Marriage, Class, and Interest: Patterns of Marriage among Pakhtun in Kunar," *Folk*, 24, pp.151–165.

Bourdieu, Pierre, 1977. *Outline of a Theory of Practice*, Cambridge University Press, Cambridge, UK.

Brickell, Chris, 2005. "Masculinities, Performativity, and Subversion: A Sociological Reappraisal," *Men and Masculinities*, 8, 1, pp.24–43.

Bromberg, Philip, 1993. "Shadow and Substance: A Relational Perspective on Clinical Process," in Bromberg, Philip, 1998. *Standing in the Spaces: Essays on Clinical Process, Trauma, and Dissociation*, pp.267–290, The Analytic Press, Hillsdale, NJ.

Bromberg, Philip, 1996. "Standing in the Spaces: The Multiplicity of the Self and the Psychoanalytic Relationship," in Bromberg, Philip, 1998. *Standing in the Spaces: Essays on Clinical Process, Trauma, and Dissociation*, pp.267–290, The Analytic Press, Hillsdale, NJ.

Bromberg, Philip, 2003. "One Need Not Be a House to Be Haunted: On Enactment, Dissociation, and the Dread of 'Not-Me'—A Case Study," *Psychoanalytic Dialogues*, 13, pp.689–709.

Bromberg, Philip, 2011. *The Shadow of the Tsunami, and the Growth of the Relational Mind*, Routledge, New York.

Bruner, Jerome, 1990. *Acts of Meaning*, Harvard University Press, Cambridge, MA.

Butler, Judith, 1988. "Performative Acts and Gender Constitution: An Essay in Phenomenology and Feminist Theory," *Theater Journal*, 40, 4, pp.519–531.

Butler, Judith, 1990. *Gender Trouble: Feminism and the Subversion of Identity*, Routledge, New York.

Butler, Judith, 1993. *Bodies That Matter: On the Discursive Limits of "Sex,"* Routledge, New York.

Butler, Judith, 1997. *The Psychic Life of Power*, Stanford University Press, Stanford, CA.

Calabrese, Joseph, 2013. *A Different Medicine: Post-Colonial Healing in the Native American Church*, Oxford University Press, New York.

Campbell, Donald, and Robert LeVine, 1961. "A Proposal for Cooperative Cross-Cultural Research on Ethnocentrism," *Journal of Conflict Resolution*, 5, pp.82–108.

Carstairs, G. M., 1958. *The Twice-Born: Study of High-Caste Community Hindus*, Indiana University Press, Bloomington, IN.

Casimir, Michael, 2010. *Growing Up in a Pastoral Society: Socialization among Pashtu Nomads in Afghanistan*, Kölner Ethnologishe Beiträge, Heft 33, Köln.

Chapin, Bambi, 2013. "Attachment in Rural Sri Lanka: The Shape of Caregiver Sensitivity, Communication, and Autonomy," in Quinn, Naomi, and Jeanette Mageo, eds., *Attachment Reconsidered: Cultural Perspectives on a Western Theory*, Palgrave Macmillan, New York, pp. 143–163.

Chianese, Domenico, 2007. *Constructions and the Analytic Field: History, Scenes, and Destiny*, Routledge, London.

Chiovenda, Andrea, 2018a: "'The War Destroyed Our Society': Masculinity, Violence and Shifting Cultural Idioms Among Afghan Pashtuns", in Shahrani, Nazif, ed. *Modern Afghanistan: The Impact of 40 Years of War*, Bloomington: Indiana University Press, pp. 179–199.

Chiovenda, Andrea 2018b. "Shaping a Different Masculinity: Afghan Pashtun Men's Untold Side", in Inhorn, Marcia and Nefissa Naguib, eds. *Reconceiving Muslim Men: Love and Marriage, Family and Care in Precarious Times*, New York: Berghahn Books, pp. 63–84.

Chiovenda, Andrea, in press. "From Metaphor to Interpretation: 'Haunting' as Diagnostic of Dissociative Processes", *Ethos*.

Chiovenda, Melissa, 2012. *Agency through Ambiguity: Women NGO Workers in Jalalabad, Afghanistan*, Master's Thesis, University of Connecticut, Storrs, CT.

Chiovenda, Melissa, 2014. "The Illumination of Marginality: How Ethnic Hazaras in Bamyan, Afghanistan, Perceive the Lack of Electricity as Discrimination," *Central Asian Survey*, 33, 4, pp.1–14.

Chiovenda, Melissa, 2015. "Memory, History, and Landscape: Ethnic Hazaras' Understanding of Marginality in Bamyan, Afghanistan," in Kukreja, Sunil, ed., *State, Society, and Minorities in South and Southeast Asia*, Lexington Books, Lanham.

Chiovenda, Andrea and Melissa Chiovenda, 2018. "The Specter of the 'Arrivant': Hauntology of an Interethnic Conflict in Afghanistan", *Asian Anthropology*, 17: 165–184.

Chivers, C. J., and Dexter Filkins, 2010. "Coalition Troops Storm a Taliban Haven," *The New York Times*, February 13, Page A1.

Civitarese, Giuseppe, 2010. *The Intimate Room: Theory and Technique of the Analytic Field*, Routledge, London.

Civitarese, Giuseppe, 2013. *The Violence of Emotions: Bion and Post-Bionian Psychoanalysis*, Routledge, London.

Connell, Robert W., 1987. *Gender and Power: Society, the Person, and Sexual Politics*, Allen and Unwin, Sydney.

Connell, Robert W., 1995. *Masculinities*, Allen and Unwin, Sydney.

Crapanzano, Vincent, 1980. *Tuhami, Portrait of a Moroccan*, University of Chicago Press, Chicago.

Dalal, Farhad, 2002. *Race, Colour, and the Process of Racialization: New Perspectives from Group Analysis, Psychoanalysis, and Sociology*, Routledge, London.

Dalal, Farhad, 2006. "Racism: Processes of Dehumanization, Detachment, and Hatred," *Psychoanalytic Quarterly*, 75, pp.131–161.

Delaney, Carol, 1986. "The Meaning of Paternity and the Virgin Birth Debate," *Man*, 21, 3, pp.494–513.

Delattre, Etienne, and Haqiq Rahmani, 2007. *A Preliminary Assessment of Forest Cover and Change in the Eastern Forest Complex of Afghanistan*, Wildlife Conservation Society and USAID, Internet publication.

Demetriou, Demetriakis, 2001. "Connell's Concept of Hegemonic Masculinity: A Critique," *Theory and Society*, 30, 3, pp.337–361.

Derrida, Jacques. 1994. *Specters of Marx: The State of the Debt, the Work of Mourning, and the New International*, New York: Routledge

Devereux, George, 1951. *Reality and Dream: Psychotherapy of a Plains Indian*, International Universities Press, New York.

Devereux, George, 1967. *From Anxiety to Method in the Behavioral Sciences*, Mouton, The Hague.

Doi, Takeo, 1981. *The Anatomy of Dependence*, Kodansha International Ltd., Tokyo.

Doi, Takeo, 1986. *The Anatomy of the Self: The Individual Versus Society*, Kodansha International Ltd., Tokyo.

Dwairy, Marwan, Mona Fayad, and Naima BenYaqoub, 2013. "Parenting Profiles versus Parenting Factors and Adolescents' Psychological Disorders," *Journal of Educational and Developmental Psychology*, 3, 2, pp.1–23.

Edwards, David, 1996. *Heroes of the Age: Moral Fault Lines on the Afghan Frontier*, University of California Press, Berkeley, CA.

Edwards, David, 1998. "Learning from the Swat Pathans: Political Leadership in Afghanistan, 1978–1997," *American Ethnologist*, 25, 4, pp.712–728.

Edwards, David, 2002. *Before the Taliban: Genealogies of the Afghan Jihad*, University of California Press, Berkeley, CA.

Elias, Norbert, 1978. *What Is Sociology?*, Columbia University Press, New York.

Elias, Norbert, 1991 [1939]. *The Society of Individuals*, Basil Blackwell, Oxford, UK.

Elias, Norbert, 1994 [1968]. "Postscript," in *The Civilizing Process*, Basil Blackwell, Oxford, UK, pp. 449–484.

Erikson, Erik, 1950. *Childhood and Society*, Norton, New York.

Ewing, Katherine, 1987. "Clinical Anthropology as an Ethnographic Tool," *Ethos*, 15, pp.16–39.

Ewing, Katherine, 1990. "The Illusion of Wholeness: Culture, Self, and the Experience of Inconsistency," *Ethos*, 18, pp.251–273.

Ewing, Katherine, 1991. "Can Psychoanalytic Theories Explain the Pakistani Woman? Intrapsychic Autonomy and Interpersonal Engagement in the Extended Family," *Ethos*, 19, pp.131–160.

Ewing, Katherine, 1992. "Is Psychoanalysis Relevant to Anthropology? ," in Schwartz, Theodore, G. White, and C Lutz, eds., *New Directions in Psychological Anthropology*, Cambridge University Press, Cambridge, UK, pp. 251–268.

Ewing, Katherine, 1997. *Arguing Sainthood: Modernity, Psychoanalysis and Islam*, Durham: Duke University Press.

Ewing, Katherine Pratt, 2008. *Stolen Honor: Stigmatizing Muslim Men in Berlin*, Stanford: Stanford University Press.

Fassin, Didier, 2000. "Les politiques de l'ethnopsychiatrie: La psyché africaine, des colonies africaines aux banlieues parisiennes," *L'Homme*, 153, pp.231–250.

Feldman, Allen, 1991. *Formations of Violence: The Narrative of the Body and Political Terror in Northern Ireland*, University of Chicago Press, Chicago.

Ferdinand, Klaus, 2006. *Afghan Nomads: Caravans, Conflicts, and Trade in Afghanistan and British India*, Copenhagen.

Ferrari, Armando, 2004. *From the Eclipse of the Body to the Dawn of Thought*, Free Association Books, London.

Ferro, Antonino, 2002. "Some Implications of Bion's Thought: The Waking Dream and Narrative Derivatives," *International Journal of Psychoanlysis*, 83, pp.597–607.

Ferro, Antonino, 2005. "Bion: Theoretical and Clinical Observations," *International Journal of Psychoanalysis*, 86, pp.1535–1542.

Ferro, Antonino, 2006a. "Trauma, Reverie, and the Field," *Psychoanalytic Quarterly*, 75, pp.1045–1066.

Ferro, Antonino, 2006b. "Clinical Implications of Bion's Thought," *International Journal of Psychoanalysis*, 87, pp.989–1003.

Ferro, Antonino, 2009. "Tranformations in Dreaming and Characters in the Psychoanalytic Field," *International Journal of Psychoanalysis*, 90, pp.209–230.

Ferro, Antonino, 2011. *Avoiding Emotions, Living Emotions*, Routledge, Harvey, UK.

Ferro, Antonino, and Roberto Basile, 2009. "The Universe of the Field and Its Inhabitants," in Ferro, Antonino, and Roberto Basile, eds., *The Analytic Field: A Clinical Concept*, Routledge, New York, pp. 5–30.

Ferro, Antonino, and Giuseppe Civitarese, 2013. "Analysts in Search of an Author: Voltaire or Artemisia Gentileschi? Commentary on 'Field Theory in Psychoanalysis, Part 2: Bionian Field Theory and Contemporary Interpersonal/Relational Psychoanalysis,'" *Psychoanalytic Dialogues*, 23, pp.646–653.

Fonagy, Peter, 2001. *Attachment Theory and Psychoanalysis*, Other Books, New York.

Freud, Sigmund, 1937. "Analysis Terminable and Interminable," *International Journal of Psychoanalysis*, 18, pp.373–405.

Frie, Roger, 2008a. "Introduction: The Situated Nature of Psychological Agency," in Frie, Roger, ed., *Psychological Agency: Theory, Practice, and Culture*, The MIT Press, Cambridge, MA, pp. 1–31.

Frie, Roger, 2008b. "Navigating Cultural Contexts: Agency and Biculturalism," in Frie, Roger, ed., *Psychological Agency: Theory, Practice, and Culture*, The MIT Press, Cambridge, MA, pp. 223–240.

Fromm, Erich, and Michael Maccoby, 1970. *Social Character in a Mexican Village: A Socio-psychoanalytic Study*, Prentice-Hall, Englewood Cliffs, NJ.

Geertz, Clifford, 1973. *The Interpretation of Cultures*, Basic Books, New York.

Geertz, Clifford, 1984. "From the Native's Point of View: On the Nature of Anthropological Understanding," in Schweder, Robert, and Robert LeVine, eds., *Culture Theory: Essays on Mind, Self, and Emotion*, Cambridge University Press, pp. 123–136.

Ghani, Ashraf, 1978. "Islam and State-Building in a Tribal Society. Afghanistan, 1880–1901," *Modern Asian Studies*, 12, 2, pp.269–284.

Ghannam, Farha, 2013. *Live and Die like a Man: Gender Dynamics in Urban Egypt*, Stanford University Press, Stanford, CA.

Ghassem-Fachandi, Parvis, ed., 2009. *Violence: Ethnographic Encounters*, Berg, Oxford, UK.

Gilmore, David. 1987. *Honor and Shame and the Unity of the Mediterranean*, American Anthropological Association, Washington, DC.

Gilmore, David, 1990. *Manhood in the Making: Cultural Concepts of Masculinity*, Yale University Press, New Haven, CT.

Giovannini, Maureen. 1981. "Woman: A Dominant Symbol within the Cultural System of a Sicilian Town," *Man*, 16, pp.408–426.

Giustozzi, Antonio, 2003. "Respectable Warlords? The Politics of State-building in Post-Taleban Afghanistan," *Working Papers no.1*, Crisis States Research Center, LSE, London.

Giustozzi, Antonio, 2006. " 'Tribes' and Warlords in Southern Afghanistan, 1980–2005," *Working Papers no.7*, Crisis States Research Center, LSE, London.

Giustozzi, Antonio, 2008. *Koran, Kalashnikov, and Laptop: The Neo-Taliban Insurgency in Afghanistan*, Hurst, London.

Giustozzi, Antonio, 2009. *Empires of Mud: War and Warlords in Afghanistan*, Columbia University Press, New York.

Glatzer, Bernt, 1977. *Nomaden von Gharjistān: Aspekte den wirtschaftlichen, sozialen, und politischen Organisation nomadischen Durrānī-Paschtunen in Nordwestafghanistan*, Steiner, Wiesbaden.

Glatzer, Bernt, 2002. "The Pashtun Tribal System," in Pfeffer, G, and D. K. Bekera, eds., 2002, Contemporary Society: *Concept of Tribal Society*, Concept Publisher, New Delhi, pp. 265–282.

Goldberg, Peter, 1995. "'Successful' Dissociation, Pseudovitality, and Inauthentic Use of the Senses," *Psychoanalytic Dialogues*, 5, pp.493–510.

Good, Byron, 2012a. "Phenomenology, Psychoanalysis, and Subjectivity in Java," *Ethos*, 40, 1, pp.24–36.

Good, Byron, 2012b. "Theorizing the 'Subject' of Medical and Psychiatric Anthropology," *Journal of the Royal Anthropological Institute*, 18, pp.515–535.

Good, Byron. 2015. "Haunted by Aceh: Specters of Violence in Post-Suharto Indonesia," in *Genocide and Mass Violence: Memory, Symptom and Recovery*, Devon E Hinton and Alexander L Hinton, eds. Pp. 58–82. Cambridge: Cambridge University Press.

Good, Byron, and Mary-Jo DelVecchio Good, 2017. "Toward a Cultural Psychology of Trauma and Trauma-Related Disorders," in Cassaniti, Julia, and Usha Menon, eds., *Universalism Without Uniformity: Explorations in Mind and Culture*, University of Chicago Press, Chicago, pp. 260–279.

Good, Byron, Henry Herrera, Mary-Jo DelVecchio Good, and James Cooper, 1982. "Reflexivity and Countertransference in a Psychiatric Cultural Consultation Clinic," *Culture, Medicine, and Psychiatry*, 6, pp.281–303.

Grima, Benedicte, 1992. *The Performance of Emotion among Paxtun Women*, University of Texas, Austin, TX.

Gutmann, Matthew, 2007 [1996]. *The Meanings of Macho: Being a Man in Mexico City*, University of California Press, Berkeley, CA.

Hallowell, Irving, 1955. *Culture and Experience*, University of Pennsylvania Press, Philadelphia.

Hammoudi, Abdellah, 1996. "Segmentarity, Social Stratification, Political Power, and Sainthood: Reflections on Gellner's Theses," in Hall, J. A., and I. Jarvie, eds., *The Social Philosophy of Ernest Gellner*, Rodopi, Amsterdam, pp. 265–289.

Hartmann, Heinz, 1958. *Ego Psychology and the Problem of Adaptation*, International Universities Press, New York.

Herzfeld, Michael, 1985. *The Poetics of Manhood: Contest and Identity in a Cretan Mountain Village*, Princeton University Press, Princeton, NJ.

Hoffman, Irwin, 1994. "Dialectical Thinking and Therapeutic Action in the Psychoanalytic Process," *Psychoanalytic Quarterly*, 63, pp.187–218.

Hollan, Douglas, 1992. "Cross-cultural Differences in the Self," *Journal of Anthropological Research*, 48, pp.283–300.

Hollan, Douglas, 2000. "Constructivist Models of Mind, Contemporary Psychoanalysis, and the Development of Culture Theory," *American Anthropologist*, 102, pp.538–550.

Hollan, Douglas, 2001. "Developments in Person-centered Ethnography," in Moore, C.C., and Matthews, H. F., eds., *The Psychology of Cultural Experience*, Cambridge University Press, Cambridge, UK, pp. 48–67.

Hollan, Douglas, 2016. "Psychoanalysis and Ethnography," *Ethos*, 44, 4, pp.507–521.

Hollan, Douglas and Jane Wellenkamp, 1994. *Contentment and Suffering:Culture and Experience in Toraja*, New York, Columbia University Press.

Hollan, Douglas and Jane Wellenkamp, 1995. *Thread of Life: Toraja Reflections on the Life Cicle*, Honolulu: University of Hawa'i Press.

Hoodfar, Homa, 2009. "Afghan Refugee Women in Iran: Revisioning the Afghan Family," in Cuno, Kenneth, and Manisha Desai, eds., *Family, Gender, and Law in a Globalizing Middle East and South Asia*, Syracuse University Press, Syracuse, NY, pp.223–247.

Inhorn, Marcia, 2012. *The New Arab Man: Emergent Masculinities, Technologies, and Islam in the Middle East*, Princeton University Press, Princeton, NJ.

Isbell, Billie Jean, 2009. "Written on My Body," in Ghassem-Fachandi, Parvis, ed., *Violence: Ethnographic Encounters*, Berg, Oxford, UK, pp. 15–34.

Jenkins, Janis, 2015. *Extraordinary Conditions: Culture and Experience in Mental Illness*, University of California Press, Oakland, CA.

Kagan, Jerome, 2013. *The Human Spark: The Science of Human Development*, Basic Books, New York.

Kakar, Hasan Kawun, 1979. *Government and Society in Afghanistan: The Reign of Amir Abd al-Rahman Khan*, University of Texas Press, Austin.

Karlstetter, Maria, 2008. *Wildlife Surveys and Wildlife Conservation in Nuristan, Afghanistan*, Wildlife Conservation Society, and USAID, Internet publication.

Khaurin, H. H., 2003. *Trees and Bushes of Afghanistan*, Food and Agriculture Organization of the United Nations.

Kinder, Donald, and Cindy Kam, 2009. *Us against Them: Ethnocentric Foundations of American Opinion*, University of Chicago Press, Chicago.

King, Diane, 2008. "The Personal Is Patrilineal: *Namus* as Sovereignty," *Identity: Global Studies in Culture Power*, 15, pp.317–342.

Kluckhohn, Clyde and Henry Murray, 1948. "Personality Formation: The Determinants", in Kluckhohn, Clyde and Henry Murray, eds., *Personality in Nature, Society and Culture*, Alfred A. Knopf, New York.

Kohut, Heinz, 1972. "Thoughts on Narcissism and Narcissistic Rage," in *Psychoanalytic Study of the Child*, 27, pp.360–400.

Kracke, Waud, 1978. *Force and Persuasion: Leadership in an Amazonian Society*, University of Chicago Press, Chicago.

Kuper, Adam, 1982. "Lineage Theory: A Critical Retrospect," *Annual Review of Anthropology*, 11, pp.71–95.

Kuper, Adam, 1999. *Culture: The Anthropologists' Account*, Harvard University Press, Cambridge, MA.

Johnson, Lyman, and Sonya Lipsett-Rivera, eds., 1998. *The Faces of Honor: Sex, Shame, and Violence in Colonial Latin America*, University of New Mexico Press, Albuquerque.

Joseph, Suad, 1999. "Introduction: Theories and Dynamics of Gender, Self, and Identity in Arab Families," in Joseph, Suad, ed., *Intimate Selving in Arab Families: Gender, Self, Identity*, Syracuse University Press, Syracuse, NY, pp. 1–17.

Lamphere, Louise, Helena Ragone, and Patricia Zavella, 1997. *Situated Lives: Gender and Culture in Everyday Life*, Routledge, New York.

Leacock, Eleanor, 1981. *Myths of Male Dominance: Collected Articles on Women Cross-Culturally*, Monthly Review Press, New York.

Leacock, Eleanor, 1983. "Interpreting the Origins of Gender Inequality: Conceptual and Historical Problems," *Dialectical Anthropology*, 7, 4, pp.263–284.

Lester, Rebecca, 2017. "Self-Governance, Psychotherapy, and the Subject of Managed Care: Internal Family Systems Therapy and the Multiple Self in a US Eating-Disorders Treatment Center," *American Ethnologist*, 44, no. 1, pp.23–35.

LeVine, Robert, 1982 [1973]. *Culture, Behavior, and Personality*, University of Chicago Press, Chicago.

LeVine, Robert, 2001. "Ethnocentrism," in *International Encyclopedia of the Social and Behavioral Sciences*, pp.4852–4854.

LeVine, Robert, and Donald Campbell, 1971. *Ethnocentrism: Theories of Conflict, Ethnic Attitudes, and Group Behavior*, Wiley, New York.

LeVine, Robert, and Karin Norman, 2001. "The Infant's Acquisition of Culture: Early Arrachment Reexamined in Anthropological Perspective," in Moore, Carmella, and Holly Mathews, eds., *The Psychology of Cultural Experience*, Cambridge University Press, Cambridge, UK, pp. 83–104.

Levitt, Peggy, 1998. "Social Remittances: Migration-Driven Local-Level Forms of Cultural Diffusion," *International Migration Review*, 32, no. 4, pp.926–948.

Levitt, Peggy, and Deepak Lamba-Nieves, 2011. "Social Remittances Revisited," *Journal of Ethnic and Migration Studies*, 37, no. 1, pp.1–22.

Levy, Robert, 1973. *Tahitians, Mind, and Experience in the Society Islands*, University of Chicago Press, Chicago.

Levy, Robert, and Douglas Hollan, 1998. "Person-centered Interviewing and Observation," in Bernard, H. R., ed., *Handbook of Methods in Cultural Anthropology*, Walnut Creek, pp. 333–364.

Lindholm, Charles, 1980. "Images of the Pathan: The Usefulness of Colonial Ethnography," in Lindholm, Charles, 1996. *Frontier Perspectives: Essays in Comparative Anthropology*, Oxford University Press, Karachi, pp. 3–16

Lindholm, Charles, 1981a. "The Structure of Violence among Swat Pukhtun," in Lindholm, Charles, 1996. *Frontier Perspectives: Essays in Comparative Anthropology*, Oxford University Press, Karachi, pp. 47–61.

Lindholm, Charles, 1981b. "History and the Heroic Pukhtun," in Lindholm, Charles, 1996. *Frontier Perspectives: Essays in Comparative Anthropology*, Oxford University Press, Karachi, pp. 62–70.

Lindholm, Charles, 1982. *Generosity and Jealousy: The Swat Pukhtun of Northern Pakistan*, Columbia University Press, New York.

Lindholm, Charles, 1986. "Leadership Categories and Social Process in Islam: The Case of Dir and Swat," in Lindholm, Charles, 1996. *Frontier Perspectives: Essays in Comparative Anthropology*, Oxford University Press, Karachi, pp. 106–120.

Lindholm, Charles, 1993. "Review of Ahmed, A., *Resistance and Control in Pakistan*," *Man*, 28, 4, pp.825–826.

Lindholm, Charles, 1996. *Frontier Perspectives*, Oxford University Press, Karachi.

Lindholm, Charles, 1997. "Does the Sociocentric Self Exist? Reflections on Markus and Kitayama 'Culture and the Self'," *Journal of Anthropological Research*, 3, pp.745–760.

Lindholm, Charles, 2008. *Culture and Authenticity*, Blackwell Publications, Oxford, UK.

Littlewood, Roland, and Maurice Lipsedge, 1989. *Aliens and Alienists: Ethnic Minorotoes and Psychiatry*, Unwin Hyman, London.

Luhrmann, Tanya, 2012. "A Hyperreal God and Modern Belief: Toward an Anthropological Theory of Mind," *Current Anthropology*, 53, 4, pp.371–394.

Lutz, Catherine, 1986. "Emotion, Thought, and Estrangement: Emotion as a Cultural Category," *Cultural Anthropology*, 1, pp.287–309.

Mageo, Jeannette, 2015. "Cultural Psychodynamics: The Audit, the Mirror, and the American Dream," *Current Anthropology*, 56, 6, pp.883–900.

Mahler, Margaret, 1979. *The Selected Papers of Margaret Mahler*, Vol. 2 "Separation-Individuation," J. Aronson Publications, New York.

Mahmood, Saba, 2005. *Politics of Piety: The Islamic Revival and the Feminist Subject*, Princeton University Press, Princeton, NJ.

Mamdani, Mahmood, 2005. *Good Muslim, Bad Muslim: America, the Cold War, and the Roots of Terror*, Three Leaves Press, New York.

Markus, H., and S. Kitayama, 1991. "Culture and the Self: Implications for Cognition, Emotion, and Motivation," *Psychological Review*, 98, pp.224–253.

Markus, H., and P. Nurius, 1986. "Possible Selves," *American Psychologist*, 41, 9, pp.954–969.

Meeker, Michael, 1976. "Meaning and Society in the Near East: Examples from the Black Sea Turks and the Levantine Arabs (1)," *International Journal of Middle East Studies*, 7, pp.243–270.

Meeker, Michael, 1980. "The Twilight of a South Asian Heroic Age: A Rereading of Barth's Study of Swat," *Man*, 15, pp.682–701.

Mikkelsen, Henrik Hvenegaard, 2016. "Unthinkable Solitude: Successful Aging in Denmark through the Lacanian Real," *Ethos*, 44, 4, pp.448–463.

Minces, Juliette, 1982. *The House of Obedience: Women in Arab Society*, Zed Press, London.

Mitchell, Stephen, 1988. *Relational Concepts in Psychoanalysis: An Integration*, Harvard University Press, Cambridge, MA.

Modell, Arnold, 1993. *The Private Self*, Harvard University Press, Cambridge, MA.

Nichols, Robert, 2008. *A History of Pashtun Migration, 1775–2006*, Oxford University Press, Karachi.

Nichols, Robert, 2013. *The Frontier Crimes Regulations: A History in Documents*, Oxford University Press, Oxford, UK.t

Nisbett, Richard, and Dov Cohen, 1996. *Culture of Honor: The Psychology of Violence in the South*, Westview Press, Boulder, CO.

Nojumi, Neamatollah, 2002. *The Rise of the Taliban in Afghanistan: Mass Mobilization, Civil War, and the Future of the Region*, Palgrave, New York.

Obeyesekere, Gananath, 1981. *Medusa's Hair: An Essay on Personal Symbols and Religious Experience*, University of Chicago Press, Chicago.

Ogden, Thomas, 1994. "The Analytic Third: Working with Intersubjective Clinical Facts," *International Journal of Psychoanalysis*, 75, pp.3–19.

Ogden, Thomas, 2004. "An Introduction to the Reading of Bion," *International Journal of Psychoanalysis*, 85, pp.285–300.

Ogden, Thomas, 2015. "Intuiting the Truth of What's Happening: On Bion's 'Notes on Memory and Desire'," *Psychoanalytic Quarterly*, 84, 2, pp.285–308.

Orange, D. M., George Atwood, and Robert Stolorow, 1997. "Intersubjectivity Theory and the Clinical Exchange," in *Working Intersubjectively: Contextualism in Psychoanalytic Practice*, The Analytic Press, Hillsdale, NJ.

Ortner, S., 1984. "Theory in Anthropology since the Sixties," *Comparative Studies in Society and History*, 26, 1, pp.126–166.

Ortner, Sherry, 1996. *Making Gender: The Politics and Erotics of Culture*, Beacon Press, Boston, MA.

Otto, Hiltrud, and Heidi Keller, 2014. *Different Faces of Attachment: Cultural Variations on a Universal Human Need*, Cambridge University Press, Cambridge, UK.

Pandey, Annerose, 2009. "Unwelcoming and Unwelcomed Encounters," in Ghassem-Fachandi, Parvis, ed., *Violence: Ethnographic Encounters*, Berg, Oxford, UK, pp. 135–144.

Parish, Steven, 1994. *Moral Knowing in a Hindu Sacred City*, Columbia University Press, New York.

Pedersen, Gorm, 1994. *Nomads in Transition: A Century of Change among the Zala Khan Khel*, London.

Peristiany, John, ed., 1966. *Honor and Shame: The Values of Mediterranean Society*, Chicago University Press, Chicago.

Quinn, Naomi, and Jeanette Mageo, eds., 2013. *Attachment Reconsidered: Cultural Perspectives on a Western Theory*, Palgrave Macmillan, New York.

Ring, Laura, 2006. *Zenana: Everyday Peace in a Karachi Apartment Building*, Indiana University Press, Bloomington, IN.

Roland, Alan, 1988. *In Search of Self in India and Japan*, Princeton University Press, Princeton, NJ.

Roland, Alan, 2011. *Journeys to Foreign Selves: Asians and Asian-Americans in a Global Era*, Oxford University Press, New Delhi.

Rosaldo, Michelle, and Louise Lamphere, eds., 1974. *Woman, Culture, and Society*, Stanford University Press, Stanford, CA.

Rubin, Barnett, 1995. *The Fragmentation of Afghanistan: State Formation and Collapse in the International System*, Yale University Press, New Haven, CT.

Runyan, Anne Sisson, and V. Spike Peterson, 2013. *Global Gender Issues in the New Millennium (4th Edition)*, Westview Press, New York.

Sachs, Wulf, 1968 [1947]. *Black Anger*, Greenwood Press, New York.

Sahlins, Marshall, 1985. *Islands of History*, University of Chicago Press, Chicago.

Sanday, Peggy, 1981. *Female Power and Male Dominance: On the Origins of Sexual Inequality*, Cambridge University Press, Cambridge, UK.

Sandler, Joseph, and Bernard Rosenblatt, 1962. "The Concept of the Representational World," *The Psychoanalytic Study of the Child*, 17, pp.128–145.

Sapir, Edward, 2002. *The Psychology of Culture: A Course of Lectures*, Mouton de Gruyter, New York.

Sharabi, Hisham, 1988. *Neopatriarchy: A Theory of Distorted Change in Arab Society*, Oxford University Press, New York.

Slavin, Malcolm, and Daniel Kriegman, 1992. *The Adaptive Design of the Human Psyche: Psychoanalysis, Evolutionary Biology, and the Therapeutic Process*, The Guilford Press, New York.

Spiro, Melford, 1951. "Culture and Personality: The Natural History of a False Dichotomy," *Psychiatry*, 14, pp.19–47.

Spiro, Melford, 1952. "Ghosts, Ifaluk, and Teleological Functionalism," *American Anthropologist*, 54, 4, 497–503.

Spiro, Melford, 1965. "Religious Systems as Culturally Constituted Defense Mechanisms," in Spiro, M., ed., *Context and Meaning in Cultural Anthropology*, New York, pp 100–113.

Spiro, Melford, 1982. "Collective Representations and Mental Representations in Religious Symbol Systems," in Fernandez, James, Melford Spiro, and Milton Singer, eds., 1982, *On Symbols in Anthropology: Essays in Honor of Harry Hoijer 1980*, Undena Publications, Malibu, CA, pp. 45–72

Stern, Daniel, 2001 [1985]. *The Interpersonal World of the Infant: A View from Psychoanalysis and Developmental Psychology*, Basic Books, New York.

Stern, Donnel B., 2010. *Partners in Thought: Working with Unformulated Experience, Dissociation, and Enactment*, Routledge, New York.

Stern, Donnel B., 2013. "Field Theory in Psychoanalysis, Part 2: Bionian Field Theory and Contemporary Interpersonal/Relational Psychoanalysis," *Psychoanalytic Dialogues*, 23, pp.630–645.

Stern, Donnel B., 2015. "The Interpersonal Field: Its Place in American Psychoanalysis," *Psychoanalytic Dialogues*, 23, pp.388–404.

Steul, Willi, 1980. *Paschtunwali—Ein Ehrenkodex und seine rechtliche Relevanz*, Steiner Verlag, Wiesbaden.

Stevens, Kara, Alex Dehgan, Maria Karlstetter, Rahmat Rawan, Muhammad Ismail Tawhid, Stephane Ostrowski, Jan Mohammad Ali, and Rita Ali, 2011. "Large Mammals Surviving Conflict in the Eastern Forests of Afghanistan," *Oryx*, 45, 2, pp.265–271.

Stolorow, Robert, 2013. "Intersubjective-Systems Theory: A Phenomenological-Contextualist Psychoanalytic Perspective," *Psychoanlytic Dialogues*, 23, pp.383–389.

Stolorow, Robert, and George Atwood, 1992. *Contexts of Being: The Intersubjective Foundations of Psychological Life*, The Analytic Press, Hillsdale, NJ.

Stolorow, Robert, and George Atwood, 1994. "The Myth of the Isolated Mind," in *Progress in Self-Psychology*, pp.233–250, Guilford Press, New York.

Stolorow, Robert, Bernard Brandschaft, and George Atwood, 1987. *Psychoanalytic Treatment: An Intersubjective Approach*, The Analytic Press, Hillsdale, NJ.

Sullivan, Harry Stack, 1953. *The Interpersonal Theory of Psychiatry*, Norton, New York.

Szabo, Albert, and Thomas Barfield, 1991. *Afghanistan: An Atlas of Indigenous Architecture*, University of Texas Press, Austin.

Tapper, Nancy, 1973. "The Advent of Pashtun Maldars in Northwestern Afghanistan," *Bulletin of the School of Oriental and African Studies*, 36, 1, pp.55–79.

Tapper, Nancy, 1977. "Pashtun Nomad Women in Afghanistan," *Asian Affairs*, 7, 2, pp.163–170.

Tapper, Nancy, 1991. *Bartered Brides: Politics, Gender, and Marriage in Afghan Tribal Society*, Cambridge University Press, Cambridge, UK.

Tapper, Richard, 1979. *Pasture and Politics: Economics, Conflict, and Ritual among Shahsevan Nomads of Northwestern Iran*, Academic Press, London.

UNODC (United Nations Office on Drugs and Crime), 2008. *Afghanistan Opium Survey 2008*, http://www.unodc.org/documents/crop-monitoring/Afghanistan_Opium_Survey_2008.pdf.

UNODCCP (United Nations Office for Drug Control and Crime Prevention), 2002. *Afghanistan Opium Survey 2002*, https://www.unodc.org/pdf/publications/afg_opium_survey_2002.pdf.

Wallace, Anthony, 1956. "Revitalization Movements," *American Anthropologist*, 58, 2, pp.264–281.

Winnicott, Donald, 1960. "Ego Distortion in Terms of True and False Self," in Winnicott, Donald, 1965. *The Maturational Process and the Facilitating Environment*, International Universities Press, New York, 140–152.

Zadran, A. S., 1977. *Socio-economic and Legal-political Processes in a Pukhtun Village, Southeastern Afghanistan*, PhD dissertation, SUNY, Buffalo.

Index